Workplace Health and Safety

Also by David Walters and Theo Nichols

WORKER REPRESENTATION AND WORKPLACE HEALTH AND SAFETY

Also by David Walters

REGULATING HEALTH AND SAFETY AT WORK: The Way Forward (*with Phil James*)
HEALTH AND SAFETY IN SMALL ENTERPRISES
REGULATING HEALTH AND SAFETY MANAGEMENT IN THE EUROPEAN UNION
WORKING SAFETY IN SMALL ENTERPRISES IN EUROPE
REGULATING HEALTH AND SAFETY AT WORK: An Agenda for Change? (*with Phil James*)
BEYOND LIMITS? Dealing with Chemical Risks at Work in Europe (*with Karola Grodzki*)
SUPPORTING HEALTH AT WORK: International Perspectives on Occupational Health Services (*with Peter Westerholm*)
WORKER REPRESENTATION AND WORKPLACE HEALTH AND SAFETY (*with Theo Nichols*)
WITHIN REACH? Managing Chemical Risks in Small Enterprises

Other books by Theo Nichols

OWNERSHIP, CONTROL AND IDEOLOGY
WORKERS DIVIDED: A Study of Shopfloor Politics (*with Peter Armstrong*)
LIVING WITH CAPITALISM: Class Relations in the Modern Factory (*with Huw Beynon*)
CAPITAL AND LABOUR: Studies in the Capitalist Labour Process
WHITE COLLAR WORKERS, TRADE UNIONS AND CLASS (*with Peter Armstrong, Bob Carter and Chris Smith*)
THE BRITISH WORKER QUESTION: A New Look at Workers and Productivity in Manufacturing Work and Occupation in Modern Turkey (*with Erol Kahveci and Nadir Sugur*)
THE SOCIOLOGY OF INDUSTRIAL INJURY: GLOBAL MANAGEMENT, LOCAL LABOUR: Turkish Workers and Modern Industry (*with Nadir Sugur*)
LABOUR IN A GLOBAL WORLD: Case Studies from the White Goods Industry in Africa, South America, East Asia and Europe (*with Surhan Cam*)
THE OTHER CAR WORKERS: Work, Organisation and Technology in the Maritime Car Carrier Industry (*with Erol Kahveci*)
THE FORDISM OF FORD AND MODERN MANAGEMENT: Fordism and Post-Fordism, 2 vols (*with Huw Beynon*)
PATTERNS OF WORK IN THE POST FORDIST ERA, FORDISM AND POST-FORDISM, 2 vols (*with Huw Beynon*)
WORKER REPRESENTATION AND WORKPLACE HEALTH AND SAFETY (*with David Walters*)

Workplace Health and Safety

International Perspectives on Worker Representation

Edited by

David Walters
Professor of Work Environment, Cardiff Work Environment Research Centre, School of Social Sciences, Cardiff University

and

Theo Nichols
Distinguished Research Professor, School of Social Sciences, Cardiff University

First published 2009 by
PALGRAVE MACMILLAN

Palgrave Macmillan in the UK is an imprint of Macmillan Publishers Limited
registered in England, company number 785998, of Houndmills, Basingstoke,
Hampshire RG21 6XS.

Palgrave Macmillan in the US is a division of St Martin's Press LLC,
175 Fifth Avenue, New York, NY 10010.

Palgrave Macmillan is the global academic imprint of the above companies
and has companies and representatives throughout the world.

Palgrave® and Macmillan® are registered trademarks in the United States,
the United Kingdom, Europe and other countries.

ISBN- 13: 978-0-230-21485-9   hardback

This book is printed on paper suitable for recycling and made from fully
managed and sustained forest sources. Logging, pulping and manufacturing
processes are expected to conform to the environmental regulations of the
country of origin.

A catalogue record for this book is available from the British Library.

Library of Congress Cataloging-in-Publication Data

Workplace health and safety : international perspectives on worker
  representation / edited by David Walters and Theo Nichols.
    p. cm.
Includes bibliographical references and index.
ISBN 978-1-349-30338-0 (alk.paper)
1. Industrial hygiene. 2. Industrial safety. 3. Industrial relations.
4. Industrial management—Employee participation. I. Walters,
David, 1950- II. Nichols, Theo.

HD7261.W68 2009
363.11—dc22   2009013640

10 9 8 7 6 5 4 3 2 1
18 17 16 15 14 13 12 11 10 09

Contents

Tables

Figures

Acknowledgements

This book developed from a seminar organised by the Cardiff Work Environment Research Centre (CWERC) and held at the School of Social Sciences, Cardiff University in October 2007. We are grateful for the financial support for this seminar, provided by the School of Social Sciences, the Health and Safety Office of the European Trades Union Institute (ETUI) and from sponsorship of CWERC by the Institution for Occupational Safety and Health (IOSH), Remploy and the Royal Mail. An earlier version of Chapter 7 appeared as 'Employee "Voice" and Working Environment in Post-Communist New Member States: An Empirical Analysis of Estonia, Latvia and Lithuania', *Industrial Relations Journal*, Vol. 39, No. 4, 2008, pp. 314–34. The assistance of the UK Data Archive (UKDA) is gratefully acknowledged for providing the WIRS data set used in Chapter 1.

We are especially grateful to Sandra Bonney for her help in producing some of the figures and tables and the final text of the present volume.

Notes on Contributors

Dace Calite is a lecturer in sociology at the University of Ventspils, Latvia, and a doctoral candidate in the Department of Sociology, University of Latvia.

Marlea Clarke is completing a postdoctoral fellowship at McMaster University, Canada. She is also working on the project reported here and has recently completed a Ph.D. on precarious employment in South Africa.

Thomas Coutrot is a senior social science researcher at Dares, Ministry of Labour, France.

Alice de Wolff has worked as a community researcher on a number of projects. Most recently she was the community co-chair of the Community University Research Alliance on Precarious Employment at York University, Canada.

Isabel Dudzinski is research manager at the Trade Union Institute for Work, Environment and Health (ISTAS – *Instituto Sindical de Trabajo, Ambiente y Salud*), Madrid, Spain.

Kaj Frick is a professor at Luleå and Mälardalen Universities, and before its closure in 2007 was a senior researcher at the National Institute for Working Life, Sweden.

Rafael Gadea is a technician and researcher at the Trade Union Institute for Work, Environment and Health (ISTAS, Valencia, Spain).

Ana Maria García is an epidemiologist, occupational health researcher and senior lecturer at the University of Valencia, Spain. She also is a technician at the Trade Union Institute for Work, Environment and Health (ISTAS, Valencia, Spain) where she mostly supports research activities.

Maria J. López-Jacob is a technician researcher at the Trade Union Institute for Work, Environment and Health (ISTAS, Madrid, Spain).

Phil James is professor of employee relations at Oxford Brookes University, UK. He has written extensively in the fields of labour relations and regulation of health and safety at work in the UK.

Richard Johnstone is a professor in the Griffith Law School, Australia, where he is the Director of the Centre for Socio-Legal Research.

Until April 2004 he was the foundation Director of the National Research Centre for Occupational Health and Safety Regulation, based in the Regulatory Institutions Network in the Research School of Social Sciences at the Australian National University.

Epp Kallaste is a doctoral student in the Faculty of Economics and Business Administration, University of Tartu, Estonia.

Wayne Lewchuk is currently a professor in the Labour Studies Program and Department of Economics at McMaster University, Canada. He is the former director of the Labour Studies Program. He has studied the impact of work organisation in automobile plants on health and is presently leading a team of researchers funded by the Workplace Safety and Insurance Board and the Lupina Foundation examining the impact of temporary employment relationships on health.

Rebecca Loudoun is a lecturer in industrial relations at Griffith University, Australia.

Theo Nichols is Distinguished Research Professor, Cardiff School of Social Sciences, UK. He was one of the first sociologists in the UK to research health and safety and has written widely on a variety of subjects in the general field of economic sociology. He is the co-author with David Walters of *Worker Representation and Workplace Health and Safety.*

Fernando Rodrigo is Director and researcher at the Trade Union Institute for Work, Environment and Health (ISTAS, Valencia, Spain).

Laurent Vogel has a Ph.D. in law from Nantes University (France) with a doctoral thesis on the impact of European Union Directives on health and safety law in European countries. He is a researcher and Head of the Health and Safety Department of the European Trade Union Institute for Research, Education Health and Safety.

David Walters is Professor of Work Environment and Director of Cardiff Work Environment Research Centre (CWERC) at Cardiff University, UK. He has undertaken many research projects on occupational health and safety management, social dialogue, national strategies on regulating risk at the workplace and on health and safety in small enterprises. He is co-author with Theo Nichols of *Worker Representation and Workplace Health and Safety.*

Charles Woolfson is Professor of Labour Studies, School of Law, University of Glasgow. Between 2004 and 2007 he held a Marie Curie Chair at the University of Latvia.

Abbreviations

ACT	Australian Capital Territory
ACTOHS statutes	Australian Capital Territory Occupational Health and Safety statutes
ACTU	Australian Council of Trade Unions
AFL-CIO	American Federation of Labor and Congress of Industrial Organizations
AIR	Australian Industrial Relations Commission
AWIRS	Australian Work Industrial Relations Survey
BSSRS	British Society for Social Responsibility in Science
BWEL	Baltic Working Environment and Labour
CATI	Computer Assisted Telephone Interview
CC.OO	*Confederación Sindical de Comisiones Obreras* (Spain)
CFDT	*Confédération française démocratique du travail*
CHSCT	*Comité d'hygiène, de sécurité et des conditions de travail*
COMAH Regulations	Control of Major Accident Hazards Regulations 1999
Cth	Commonwealth (Australia)
CWERC	Cardiff Work Environment Research Centre
DG	Directorate General (European Commission)
DTI	Department of Trade and Industry (UK)
DWG	Designated work group (Australia)
ECJ	European Court of Justice
EEC	European Economic Communities
EU	European Union
GDP	Gross Domestic Product
GNP	Gross National Product
HRM	Human Resource Management
HSCE Regulations	Health and Safety (Consultation with Employees) Regulations 1996 (UK)
HSC	Health and Safety Commission (UK)
HSC	Health and Safety Committee (Australia, France and elsewhere)
HSE	Health and Safety Executive

HSR	Health and Safety Representative (Australia)
HSW Act	Health and Safety at Work etc Act 1974 (UK)
ICE Regulations	Information and Consultation Regulations (UK 2004)
ILO	International Labour Organisation
IOSH	Institution for Occupational Safety and Health
ISO	International Organisation for Standardisation
ISTAS	*Instituto Sindical de Trabajo, Ambiente y Salud*
INSEE	*Institut national de la statistique et des études économiques* (National Institute for Statistics and Economic Studies – France)
LO	*Landsorganisationen I Sverige* (The main trade union confederation in Sweden)
NACE codes	*Nomenclature Generale des Activites Economiques dans l'Union Europeenne* (System for classifying economic activities in the European Union)
NMS	New Member States (of European Union)
NSW	New South Wales
OHS	Occupational Health and Safety
OHSA	Occupational Health and Safety Act (Australia)
OHSA (ACT)	Occupational Health and Safety Act (Australian Capital Territory)
OHSA (Cth)	Occupational Health and Safety Act (Commonwealth)
OHSA (Vic)	Occupational Health and Safety Act (Victoria, Australia)
OHSA (WA)	Occupational Health and Safety Act (Western Australia)
OHSWA(SA)	Occupational Health, Safety and Welfare Act (South Australia)
PIN	Provisional Improvement Notice (most Australian States)
PPE	Personal protective equipment
Qld	Queensland, Australia
REACH	Registration, Evaluation and Authorisation of Chemicals Regulations (EU)
REPONSE Survey	*Relations Professionnelles et Negociations d'Entreprise* (National survey of employment relations – France)

SACO	*Sveriges akademikers centralorganisation* (Confederation of Professional Associations)
SEK	Swedish kronor (Swedish currency unit)
SME	Small- and Medium-Sized Enterprise
SRSC Regulations	Safety Representatives and Safety Committees Regulations 1997
SUMER Survey	*Surveillance Medicale des Risques Professionnels* (National survey of occupational risks – France)
TCO	The Swedish Confederation for Professional Employees
TUC	Trade Union Congress (UK)
UGT	*Union General de Trabajadores* (Spain)
VDU	Visual Display Unit
VTHC	Victoria Trades Hall Council (Australia)
WA	Western Australia
WERS	Workplace Employee (Industrial) Relations Survey (previously WIRS)
WHMIS	Workplace Hazardous Materials Information System (Canada)
WHA (NT)	Workplace Health Act (Northern Territory)
WHSA	Workplace Health and Safety Act
WHSA (Qld)	Workplace Health and Safety Act (Queensland, Australia)

Introduction: Representing Workers on Health and Safety in the Modern World of Work

David Walters and Theo Nichols

Representing workers on health and safety in the modern world of work

It is estimated that in the UK alone it there are over 20,000 deaths each year that are attributable to ill health or injuries related to work. In addition, more than 25,000 people leave employment as a result of work-related injury or illness, and more than two million suffer from ill health which in their view was caused or made worse by their work, leading to the loss of more than 30 million working days each year (James and Walters, 2005). Given that the UK is apparently a comparatively good performer according to analysis of the experience of work-related injuries and fatalities in the European Union (EU), this suggests that the scale of this harm is proportionally even higher in most other countries. This represents a massive burden of human suffering; but there is no great mystery surrounding its origin. It is primarily the result of failures of management in the discharge of employers' legal duties to protect the health and safety of their workers, and as such it is largely preventable.

It is for these reasons there are moral and ethical grounds for arguing that people at work require rights to representation of their health and safety interests. They need them both to ensure a degree of ability to protect their health and safety from exploitation in the name of profit (which, to managers, may sometimes appear as the rational pursuit of efficiency) and to oblige employers to make better use of their collective knowledge and experience to help improve health and safety management. It is, therefore, no coincidence or surprise that research shows that, provided certain preconditions apply, when these rights are enacted through arrangements for joint consultation at the workplace,

1

health and safety performance is better than when it is managed by employers in the absence of such arrangements (Nichols, Walters and Tasiran, 2007; Walters and Nichols, 2007). However, the key qualification to this is found in the words 'provided certain preconditions apply' because, as the same research in the UK has also demonstrated, such preconditions are by no means universally found.

It is important, therefore, to understand what is meant by these generalisations in more detail and to document what we know from research findings about the practices they describe across the range of countries in which they are found. It is also necessary to be clear that what is being discussed here is workers' representation and to distinguish this from vague notions of 'consultation', 'engagement', and 'participation' that are often used in the same context but with very different connotations.[1] 'Workers' representation' refers to the situation in which workers represent their interests through the normal channels of labour relations. While the predominant approach to achieve this is through a regulatory framework that legitimises workplace institutions that give workers a voice on matters of their health and safety, the process of effective representation always occurs through channels of workplace labour relations, is supported by organised labour and is subject to many of the same influences on the nature and extent of these wider relations.

This book brings together research that reviews the coverage and effectiveness of arrangements for worker representation on health and safety across a range of countries. It examines how these arrangements deliver results in different countries in which broadly similar approaches have been adopted towards framing the regulatory provisions that govern these matters in different sectors and organisational sizes. It also considers in some detail the conditions that support or constrain these arrangements in different countries and especially the impact upon them of major political and economic changes in recent years. Here we explain the aims of the book and the way its content is structured, and say something of the background to the issues covered.

History

The predominant model of worker representation on health and safety is located within a broadly similar regulatory context in the countries covered in the following chapters, although its historical origins in each country may be markedly different. That is, it is largely based on regulations, introduced across a range of countries more or less in the same period and consisting of broadly similar requirements. It would be easy to identify these as those embraced within the occupational health and safety requirements of ILO Convention 155 – which lays down a basic framework for workers' representation on health and safety – and therefore as products of its influence. However, we believe such an interpretation to be both simplistic and misguided. While there is resemblance between national requirements and those of the Convention, the underlying influences that they reflect are both older and more powerful – at least as far as the countries addressed here are concerned. Indeed, Convention 155, which was not adopted until 1981, is itself arguably more likely to be a product of these same influences rather than their cause.

The underlying model that has determined the inclusion of the countries covered here is essentially that of an Anglo–European approach to regulating health and safety management introduced during the decades since the 1970s – first in Scandinavian countries and then the UK, before spreading to other English-speaking countries like Australia, New Zealand and Canada while at the same time taking hold in western Europe, influencing legislative content at the level of the EU and as a consequence being widely implemented in member states. Notably absent from this list is the United States. The basic differences between the regulatory approach in which worker representation is arguably one of the cornerstones of wider arrangements for risk management such as is found in Europe and the other countries included here on the one hand and the more unilateral model that has underpinned US health and safety regulation since the introduction of the Occupational Health and Safety Act in the early 1970s on the other has led us to exclude the US experience from our coverage while recognising its importance. In fact, it is arguable that social scientific interest in worker representation in health and safety at work remains undeveloped in the US; and although there is growing interest in the emergence of workers' centres these are not situated in the workplace and stand outside the regular labour relations system (Fine, 2005). Of course, neither the regulatory frameworks for worker representation nor their economic and labour relations contexts are identical in the countries covered. But it is this diversity that makes for a rich source of comparison between the countries included, from which conclusions concerning the supports for and constraints on good practice may be drawn.

The several significant influences on the regulatory model for worker representation on health and safety in the countries concerned can be traced to wider developments in health and safety regulation, the role of organised labour and the climate of labour relations. For example, one reason for the development of the regulatory model is found in the shift from prescriptive to process-based regulation that has been evident internationally since the 1970s. This entailed a general trend away from the proliferation of detailed regulatory duties on health and safety to requirements for more general standards of health and safety management. This development was to some degree already present in Scandinavian countries from the 1960s, but it was given its defining features in the UK in the 1970s when, as a result of the recommendations of the Robens Committee, the Health and Safety at Work (HSW) Act 1974 superimposed a form of 'regulated self-regulation' on the large number of existing statutory arrangements for the protection of workers' health and safety that had been established by successive Factories Acts and related legislation since the early 19th century.

While they may have had different legal antecedents, other countries in northern Europe such as the Netherlands, as well as those with predominantly English-based legal constitutions such as Australia, Canada and New Zealand were quick to follow suit, and the international influence of the Robens Report is widely acknowledged in this respect (Gunningham and Johnstone, 1999). As we shall discuss in more detail later, in this shift towards more process-based regulation of health and safety management there was a presumption, widely held, of the need to engage workers and or their representatives in the self-regulatory processes with which employers were expected to discharge their duties.

The second important influence that has helped to embed these notions across an even broader range of countries was provided by the drive towards harmonisation of the regulation of health and safety within the EU and in particular by the measures on systematic health and safety management adopted in the EU Framework Directive 89/391. This Directive built its focus on employers' duties to manage health and safety on several key concepts. They include the need to assess workplace risks and to respond to their management with reference to a hierarchy of control preferences that were to be implemented by following competent advice and in consultation with workers and/or their representatives (Walters, 2002). The result was that the harmonising effects of the Framework Directive brought about the adoption of a broadly similar model of process-based regulation of health and safety management in which provision for consultation and representation of workers is a significant feature in countries with somewhat different legal and industrial relations traditions and practices.

A third set of influences on the development of the regulatory model for worker representation was the role of organised labour and the broader labour relations climate. These influences are more complex. Nevertheless, they are important underpinnings of the regulatory model and help to explain both its content and its dominant international position. One national and one international example will suffice to illustrate these points here – although others feature in subsequent chapters. The British case provides a national example of the influence of organised labour on the content of regulation, both in terms of its impact and in terms of the position labour held in the political economy at the time, which made such an impact possible. As described elsewhere, the watershed for regulatory measures on worker representation in the UK occurred during the 1970s – when trade unions were in a position of maximum strength and political influence – with the introduction of the Safety Representatives and Safety Committees (SRSC) Regulations 1977 under the HSW Act. The regulations were a product of a long trade union campaign that is traceable to the 1950s (Grayson and Goddard, 1975; Walters, 1996). They may have been made in the spirit of the Robens-influenced HSW Act but, in providing trades unions with exclusive rights to appoint health and safety representatives, they were a conscious departure from the unitarist industrial relations thinking that dominated the Robens approach and the product of a quite separate trajectory in which the political power that organised labour was able to wield at the zenith of the post-war compromise was instrumental. In this respect, as Walters (1996, 2006) has previously demonstrated, it was the political influence of the trade union campaign, and in particular the wider agreement between the trade unions and the 1974–8 Labour Government (the so-called Social Contract), that allowed the unions to obtain the wording to the regulations that best suited their purpose – transforming a general requirement to consult workers that was the recommendation of the Robens Report and the first Health and Safety at Work Bill into a specific requirement to grant trades unions (and only trades unions) rights to appoint health and safety representatives and to enable these representatives to require employers to establish joint health and safety committees.

The second example of the role of labour is found in the compromises that led to adoption of the wording of the EU Framework Directive 89/391 in 1989. As both Vogel (1994) and Walters (2002) have separately documented, the final content of this Directive was very much the consequence of a compromise brought about by the Commission's need to press forward with the adoption of the Machinery Directive.

This Directive was linked to the drive towards removing barriers to trade. The Framework Directive was linked to the introduction of minimum standards to protect workers – the so-called social dimension of EU employment policy at the time. The need for harmony between these two strands of Commission policy – freeing up trade and protecting labour – was regarded as an essential feature of EU policy on trade and employment overall, and this allowed the trade unions in the Advisory Committee on Safety, Health and Hygiene an opportunity to achieve an advantage they were able to exploit in pressing for general requirements on consultation with 'workers and their representatives' to be included in the wording of the Directive. That this occurred against the wishes of the employers' organisations at the European level, despite the shift that had by then occurred in the balance of power from labour towards capital in EU countries and at the level of the EU, and also at a time when the political strategies of some of the more powerful member states (such as the UK, for example) were actively hostile to the trades unions, is remarkable. It suggests that decisions at the EU level are a function of both power and opportunity and cannot always be read off in a seemingly straightforward way.

The model

Whatever the reasons that lie behind national regulatory models on worker representation and those of the EU, there is broad similarity in their requirements. As Walters and Nichols (2007, p. 13) describe, they include:

- selection of health and safety representatives by employees;
- protection from victimisation or discrimination as a result of the representative role;
- paid time off to carry out the function of a safety representative;
- paid time off for training to function as a safety representative;
- rights to receive information from employers on hazards to the health and safety of workers at the workplace;
- the right to inspect the workplace, investigate complaints from workers on health and safety matters and to make representations to the employer on these matters;
- the right to be consulted over health and safety arrangements, including future plans;
- the right to be consulted about the use of specialists in health and safety by the employer; and

- the right to accompany health and safety authority inspectors when they inspect the workplace and to make complaints to them when necessary.

There are, of course, variations in the details of these arrangements from country to country, and in certain cases national provisions develop the model somewhat further. For example, in Sweden there is provision for regional health and safety representatives; and in Australia health and safety representatives are allowed to issue Provisional Improvement Notices and also to grant trade union officials rights of access to workplaces. The significance of these particular provisions will be addressed in detail in subsequent chapters.

An important distinction that can be made between UK provisions and the majority of national and EU provisions concerns the role granted to organised labour. In most cases there is no explicit mention of trades unions – representation and consultation is required with 'workers' representatives' or with 'workers and/or their representatives'. As already noted, in the UK the trade unions won an exclusive right to represent workers in the SRSC Regulations. In Australia, too, the original measures on workers' representation on health and safety were more closely linked to trade union representation than is currently the case. And even in the UK, while arguably the 'preferred model' remains that of trade union representation, the Health and Safety (Consultation with Employees) Regulations 1996, introduced by a Conservative Government that was anxious to avoid falling foul of the European Court of Justice, allow for employee representation with non-union representatives as well as for consultation directly with employees, and leave it largely to the discretion of the employer which to choose. What these developments would seem to imply for these countries is a retreat from the regulatory endorsement of trade unions as central to the representation of employee interests in health and safety. Given the evidence of the role of trade unions in making such representation effective, this retreat is addressed in subsequent chapters, both specifically in relation to these countries and in relation to the role of organised labour more generally.

Different experiences of worker representation

Since the 1970s the nature of the political economy, the structure and organisation of work and the balance of power between labour and capital have all altered profoundly in the advanced market economies represented in this book. This leads to one of the central questions addressed in the following chapters: How has the model for worker representation on health and safety fared in the face of these changes?

To undertake this examination the book has been constructed in two parts. Part I outlines arrangements for workers' representation in several countries. It starts from the premise that these arrangements, which are found widely in advanced market economies, are supported by a mixture of legislative rights and industrial relations structures. These are discussed in a range of different national contexts, and their limitations in the modern world of work are identified. The research from the UK, Australia, France and Spain shows certain preconditions need to be met if worker representation and consultation is to be effective; and Part I explores the extent to which these preconditions are eroded or enhanced by current regulatory and organisational constructs in different situations across a range of countries and at the international level.

In Chapter 1 Theo Nichols and David Walters review the changes in British health and safety arrangements for formal worker representation since the 1980s. They consider the effectiveness of various arrangements, statistically and by means of recently completed case studies. They find arrangements for worker representation lacking even in the sectors in which the preconditions for effective operation might be considered the most likely to exist. They argue that this suggests a case for increasing the role of regulatory agencies in ensuring compliance with the legal requirements for representation and consultation in the UK. However, as the chapter makes clear, far from doing so, in recent years the policy on regulatory intervention on this subject in the UK has in fact moved in the opposite direction.

Somewhat in contrast to the British case, Richard Johnstone demonstrates that in Australia there has been some limited regulatory effort to address perceived challenges of change in the structure and organisation of work for the statutorily based arrangements on worker representation and consultation on workplace health and safety. In Chapter 2 he examines the statutory provisions governing health and safety representatives and committees in Australia. He identifies a dominant model for health and safety representatives based on that in operation in the UK, but with significant variations which are particularly evident in the enforcement powers vested in elected health and safety representatives. The chapter briefly notes that three Australian jurisdictions take approaches to worker representation that deviate in significant ways from the dominant Australian model. It also discusses: the general withdrawal of the privileged position of trade unions in institutional arrangements; attempts to modify institutional arrangements to enable precarious and contingent workers to be represented at workplaces; and some of the recent research data on the operation of Australian institutional arrangements for worker representation in relation to occupational health and safety.

Returning to Europe for the remaining contributions to Part 1, we first consider evidence from recent surveys of the role of joint consultation on health and safety from two EU countries, France and Spain.

In France, all firms with above 50 employees in the private sector are required by law to establish a *Comité d'hygiène, de sécurité et des conditions de travail* (CHSCT), which includes elected workers' representatives who must be consulted by the employer about health and safety issues. Thomas Coutrot presents the first national empirical study of the influence of CHSCT presence on health and safety issues at the firm level in France. It describes the background to the introduction of the requirements of CHSCTs and the spread of CHSCTs across the sectors covered, and identifies the features of workplaces in which they are most likely to be present. It further discusses evidence of their influence on strategies to reduce occupational risks in firms.

In Spain the Law for the Prevention of Occupational Health Risks (1995) requires all workplaces with six or more workers to have one or more safety representatives. Having first outlined these regulatory provisions and their industrial relations contexts in Spain, Ana Maria Garcia and her colleagues present the results of a recent study of the experiences of safety representatives. They survey the activities of health and safety representatives, their perceptions of occupational risks and their management as well as their training needs and the obstacles and supports they perceive to performing their duties. The survey represents the first systematic investigation of such issues in Spain. It shows that, while safety representatives are active in a number of ways in relation to health and safety, the extent to which they are consulted on health and safety arrangements remains limited in many workplaces, and that a significant number of safety representatives report that some of the activities for occupational health and safety protection in their companies that are required by law (such as risk evaluation, planning of preventive interventions, and workers consultation and participation in evaluation and planning) are not properly implemented.

Most of published research on representing workers on health and safety has been based on studies conducted in English-speaking countries such as the UK and Australia. While the broad legislative requirements may be similar, industrial relations cultures and contexts vary enormously between countries, and make the two surveys from France and Spain that are described in Chapters 3 and 4 particularly important in aiding a wider understanding of the issues and challenges confronting the implementation and operation of the regulatory model across a range of different national settings.

Since the EU has played such an important role in promulgating the regulatory model that is at the heart of current arrangements in the countries covered in this book, it is appropriate to end Part I with some reflections on EU policies on worker representation on health and safety. In Chapter 5 Laurent Vogel and David Walters point out that, although there are no systematic statistical data, it is estimated that there are over a million safety representatives in the European Union. They observe that, although most are trade union members, whose presence is based on broadly similar regulatory frameworks, different industrial relations systems in different member states make for widely differing styles of implementation. They offer a critique of current EU policies on employee representation at the workplace set in the context of its historical development from the 1970s onwards. They draw attention to the significance of the role of the Framework Directive 89/391 and point to some of its consequences. They further discuss a range of features of the current profile of worker representation on health and safety in the EU and the political and policy background to the challenges it faces.

These challenges set the scene for the second part of the book, which provides case studies from several further countries both within and outside the EU, and which focuses more explicitly on the challenges to worker representation presented by changes in the structure and organisation of work.

Changing times – some challenges, solutions and implications

Part II examines contemporary research findings on the challenges of precarious employment in Canada; the position of workers' representation on health and safety in the neo-liberal market economies of the former Communist Baltic states; and Swedish approaches to representing workers in small enterprises. It also presents a comparative discussion of trade union approaches to supporting worker representation in health and safety in Australia and the UK, and the extent to which such representation is embraced in union strategies to achieve renewal in these countries.

In Chapter 6 Wayne Lewchuck and his colleagues draw on a recent survey of workers in Ontario to focus on Canadian experiences of representing workers on health and safety in precarious employment. They describe the transition to an internal responsibility framework to regulate health and safety issues that took place in most Canadian jurisdictions in the late 1970s. At that time, legislation mandated Joint Health and Safety Committees in most workplaces; worker knowledge of the hazards they faced was enhanced through new right-to-know regulations; and workers were given the right to refuse dangerous work. Very little has changed in this regulatory framework in the ensuing three decades, but the effectiveness of the regulations in improving health and safety outcomes has been, and continues to be, debated. In earlier work Lewchuk argued that their effectiveness was sensitive to the labour–management environment of individual workplaces. In particular, the framework was more effective where labour was organised and where management had accepted a philosophy of co-management of the health and safety function.

While the regulatory framework has remained essentially in place, the economic and social context in which it functions has changed. Today, the economic climate has become more competitive as the Canadian economy is more reliant on foreign markets for its products. Union density has fallen, and the unions that remain face a more aggressive management focused on profitability. There have been significant changes in the structure of employment relationships. 'Employees' represent a smaller component of the total workforce, and a smaller share of those who are still classed as such is in full-time permanent employment. Non-standard forms of employment, some of it of a precarious nature, have grown in relative importance. The chapter engages with these developments and presents new evidence on the impact of some of these changes on worker voice and the nature of worker citizenship. The authors argue that they raise serious questions regarding the continuing efficacy of the internal responsibility system as a strategy for improving the health of workers.

In Chapter 7 Charles Woolfson and his colleagues examine employee voice in workplace health and safety after European Union enlargement using data gathered by means of the Baltic Working Environment and Labour (BWEL) survey. The Baltic New Member States of Estonia, Latvia and Lithuania joined the EU in 2004. Here, as elsewhere in the New Member States, the adoption of new legislative frameworks which provide for employee participation in occupational health and safety processes is often seen as evidence of the spread of a European 'social dimension'. However, the authors' survey evidence points to low levels of awareness of both mandated and union-based channels for employee voice in working environment issues. These findings are to be seen against a background of poor safety performance in the Baltic States and, in particular, evidence of work intensification. It would appear that overall problems of securing effective workplace social dialogue in the neo-liberal New Member States are also affecting occupational health and safety participation. This, the authors suggest, means that the new EU strategy (2007–12) on working environment in which social dialogue as a participative pathway to health and safety improvements is ignored, and fails to address the fundamental problem in securing employee voice in post-Communist New Member States.

To turn to successful strategies for addressing the challenges to worker representation, in Chapter 8 Kaj Frick examines a Swedish scheme to represent the health and safety interests of workers in small enterprises. One of the widely recognised challenges to institutional arrangements for representing workers in workplace health and safety concerns workplace size. In virtually all surveys of the coverage of these arrangements, representation is shown to be increasingly difficult in smaller workplaces. In this respect regional safety representatives have been a success story in Sweden since the mid-1970s. Regional safety representatives now cover some four-fifths of small workplaces with employees in Sweden, and such representatives visit workplaces between five and ten times more often than labour inspectors and considerably more than the occupational health and safety services which are also found in Sweden. They support the small worksites by appointing, training and supporting local representatives, checking on occupational health and safety problems and advising on how to solve them. Frick argues that in many cases they have resolved management issues in relation to serious risks which employers could have resolved themselves but didn't. He suggests that the level of conflict these representatives experience with employers is relatively low, and many more small employers/managers ask for their help than complain about them. He argues that the preconditions for this success have included a positive industrial relations and occupational health and safety climate, in which employers and managers accept and even appreciate cooperation with unions and their representatives – which has extended to those in small firms – and in which good occupational health and safety has significant political value.

However, Frick also suggests that the strong support for occupational health and safety in small sites that has been provided by the regional safety representatives may now be threatened by wider changes in the governance of the scheme and especially by the withdrawal of state financial support. Trends in the continued restructuring and globalisation of Swedish production may also make it harder for such representatives to reach small workplaces, which, in turn, may have fewer internal occupational health and safety resources than before, contributing further barriers to the current and future effectiveness of the scheme.

In Chapter 9 Rebecca Loudoun and David Walters examine the attitudes and perceptions of trade union officials towards health and safety representation in Britain and Australia in the light of the challenges confronting worker representation and the significant role trades unions play in supporting it. They first review the traditional roles performed by unions to support health and safety representatives, highlighting the importance of external trade union support for effective worker representation and consultation in health and safety. They summarise the well-documented difficulties that unions face in providing this support before presenting their own findings, which are based on responses from full-time trade union officials in the UK and Australia. They focus especially on how these officials perceive the nature of 'health and safety' and its links to other issues in which trade unions are interested such as work organisation. They demonstrate the existence of both similarities and differences between British and Australian perceptions and outline possible reasons for them before going on to explore attitudes among British trade union officials at national and regional levels towards the relationship between health and safety and trade union organising. They discuss the consequences of these understandings for union health and safety strategies in the context of wider aspects of trade union organising.

Lessons of the past and messages for the future

In the final chapter Phil James draws the two parts of the book together with some reflections on the problems facing worker representation in health and safety internationally, the reasons for them and the strategies so far developed to address them. Drawing on the analysis in preceding chapters, he reflects on the industrial relations and regulatory issues raised in the book. His discussion acknowledges the positive impact that worker's participation has had on health and safety activities at the workplace level, as identified in the chapters on the UK, Australia,

Canada, France and Spain, as well as the important role in this that is played by trade unions in these countries. He notes the evidence from Sweden that trade unions can have a significant impact on representing workers' interests in health and safety in difficult-to-reach contexts such as those found in small enterprises. At the same time, he also identifies the existence of major challenges to the development and maintenance of support for worker representation that are also found in these chapters – and most of all in those chapters that address precarious employment and the barriers to workers' voice present in extreme neo-liberal economies.

These reflections on international experiences nevertheless identify some common ground, both in terms of the preconditions for effective representation and in the difficulties of achieving them in current economic and political contexts. The final chapter therefore examines such similarities and contrasts to explore possible ways forward for worker representation on health and safety in the light of the current structure and organisation of work and its wider political, regulatory and economic determinants.

Wider relevance

Despite a two hundred year history of regulatory action on health and safety at work, the representation of workers' interests in these matters has attracted scant attention until relatively recently. Modern regulatory measures on workers' representation are essentially the product of the last 30 years or so, although in some countries there were legislative requirements that were much older – for example, dating from 1947 in France and the 1930s in Sweden. There were also requirements specific to particular high-risk sectors, notably in mining, in which legal arrangements for the representation of miners' interests in their health and safety have a history which in some countries dates from the early part of the 20th century. None of these older requirements has ever attracted much in the way of commentary, and little is known of their operation. In many ways this is unsurprising since traditional discourse on occupational safety and health has mostly concerned engineering, technical or legal matters in relation to safety and scientific or medical ones in relation to health.

What is perhaps more surprising is that the modern situation of workers' representation, although it is more written about than previously, remains very much outside the mainstream of academic discourse in disciplines in which the issues it confronts are of central relevance.

For example, changes that affect the way in which work is carried out have been brought about through the reassertion of the power of capital and are reflected, for example, in falling union density, greater reliance on foreign markets, more aggressive managements and significant changes in the structure of employment relationships, in which non-standard forms of employment have sometimes grown in relative importance. Such changes are much discussed by scholars of the sociology of work, industrial relations and human resource management. But seldom does this discussion focus, as Lewchuck and his colleagues do in Chapter 6, on the implications of these changes for the representation of workers' interests in their health, safety and well-being. Yet, as Lewchuk and his colleagues show, the effects can be important, and their analysis contributes significantly to the evidence base necessary to understand the implications of these changes on worker voice and the nature of worker citizenship more widely.

Evaluating the nature and impact of union-organising strategies on trade union renewal in the current economic and political environment is a major preoccupation of the recent industrial relations literature. Yet in the plethora of publications on these topics little mention has been made of the role of occupational health and safety. This is despite the large number of workers representatives that exist in the EU, the majority of whom are trade unionists involved with health and safety; and as surveys of workers themselves attest, the subject is regarded as one of the most legitimate arenas for trade union representation. Loudoun and Walters demonstrate some of the supports for and constraints on the role of health and safety representatives in trade union renewal in Chapter 9; and as Phil James notes in the final chapter, this is but a small step in an area where more work is required.

Moving beyond the discipline of industrial relations, Woolfson and his colleagues use the case of the limitations of workers' voice on occupational health and safety in the Baltic States as the tangible focus for their examination of the wider problem of securing effective workplace social dialogue in the New Member States of the EU. Again, the picture they paint takes in a much wider canvass than that solely that of health and safety, one that is familiar to scholars of comparative law and economic policy in the EU; but it is this critical focus on health and safety and workers' representation that makes their account so powerfully illustrative of the nature of the wider problems in the relationship between EU policies on social dialogue and the extreme neo-liberalism of the economic strategies in these post-Communist New Member States. Their analysis, as well as that presented in other chapters, such as those by Nichols and Walters, Johnstone, Vogel and James, is also relevant to the discussion of the meaning of regulation and the role it plays in the economic affairs of the state and of organisations. Such discussion lies at the heart of discourse on compliance in regulatory and socio-legal studies, and it also features in policy analysis, yet writers in these disciplines have largely ignored the kind of law represented by the regulatory framework for worker representation and its implications for achieving compliance – such as are discussed here.

These examples provide a strong reason to question the idea that health and safety at work is a narrow technical subject best left to scientists, doctors and engineers. They suggest that by neglecting the case of workers' representation on health and safety scholars of the sociology of work, law, management and industrial relations, among others, miss a significant opportunity to study the consequences of shifts in the balance of power in the interplay between capital, labour and the state under advanced capitalism. They show that, especially in the case of representing workers' interests, health, safety and well-being are pertinent to current concerns about the consequences of globalisation, economic liberalism, and the reorganisation and restructuring of work in advanced market economies. It is to be hoped that this is of sufficient interest to scholars from the wider disciplines involved in studying these issues to encourage them to further explore the issues raised in the following chapters. Such heightened interest would be timely. For whereas many commentators, often supported by economists, have been eager to criticise the alleged harmful effects of workers' power, arguments that point to the quite literally harmful consequences for workers' health and safety that can stem from the lack of this power are in very short supply, as are discussions about the greater need for external regulation in such circumstances.

Notes

1. See Walters and Nichols (2007: 11–18) and Walters and Frick (2000) for a fuller discussion of these distinctions.

Part I National Arrangements for Workers' Representation: Case Studies from Europe and Australia

Arrangements for worker representation on health and safety are found widely in advanced market economies, supported by a mixture of legislative rights and industrial relations structures. Part I presents a series of case studies that examine these arrangements in several countries in order to help our understanding of what makes them effective. It demonstrates that employee representation and consultation requires certain preconditions for its effectiveness, and it explores the extent to which such preconditions are challenged by the features of the restructuring of work in modern market economies. The countries studied in depth include the UK, Australia, France and Spain, and, since the EU has played an important role in promulgating the regulatory model at the heart of current arrangements in most of these countries, Part I ends with some reflections on policy and practice at this level.

Part I National Arrangements for Workers' Representation: Case Studies from Europe and Australia

1

Worker Representation on Health and Safety in the UK – Problems with the Preferred Model and Beyond

Theo Nichols and David Walters

Trade unions have played a vital part in the way in which health and safety arrangements for worker representation have worked in the UK. Most importantly, the 1977 SRSC Regulations conferred certain rights exclusively on trade union representatives, including the right to call for the establishment of a joint health and safety committee. In British legal thinking this arrangement became established as a default position or, as we call it here, a 'preferred model'; other arrangements for employees who were not in workplaces where unions were recognised thus required new arrangements to be introduced. Of course, it was not inevitable that such new provision had to rest on a non-trade union base, though this was to prove the case. For example, the rights of employees to be consulted on health and safety could have been extended to non-unionised workplaces through the auspices of trade unions that operated elsewhere, or low levels of union membership could have triggered representational rights in the absence of union recognition. Such options have not been on the political agenda in the UK, however. Meanwhile, as trade union membership has fallen, UK governments have been under pressure to comply with EU Directives. As a result of this, UK provision for health and safety consultation and representation has become complex.

Health and safety arrangements in the UK

There is no one consistent time series that permits the tracking of what has happened to health and safety arrangements for employee consultation and representation on health and safety in the UK. Fortunately, though, between 1980 and 2004 five surveys were conducted as part of the Workplace Industrial Relations Survey (later Workplace Employee

Relations Survey, WERS). These surveys collected information on health and safety arrangements in British industry, albeit with some differences in the minimum threshold of establishments surveyed. Broadly comparable information exists from these surveys for the period 1980–98, which relates to three types of arrangement whereby employees had a formal voice in health and safety together with a further residual category of 'other arrangements'. The three formal arrangements are those for joint committees which deal exclusively with health and safety matters; for joint committees that deal with health and safety along with other matters; and for cases where individual safety representatives are present but in the absence of a committee. Figure 1.1 shows that, although joint committees (we present data for the dedicated and general types combined) were less in evidence in 1998 than they had been in 1980, there was no clear trend. Even by 1998 there was likewise no clear sign of the effect of new regulations that were introduced in 1996. These, the Health and Safety (Consultation with Employees) (HSCE) Regulations, were part of a wider response by the UK to a ruling of the European Court of Justice which had overturned UK law's traditional reliance on recognised trade unions as the 'single channel' of employee representation and which held that to exclude workers for whom no union was recognised was to deprive them of their Community law right to have their representative informed and consulted (Davies and Kilpatrick, 2004: 122). The Regulations placed an obligation on employers to consult employees not covered by trade

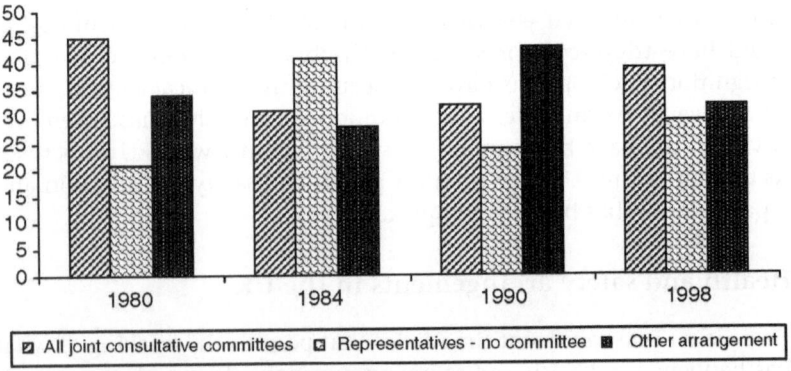

Note: Workplaces with 25 or more employees
Source: Adapted from Millward et al. (2000: 117, Figure 4.1)

Figure 1.1   Health and safety representation, 1980–98

Table 1.1 Health and safety arrangements, 1998–2004

Percentages	1998	2004
Single or multi-issue joint committees	26	20
Free-standing worker representatives	25	22
Direct methods	47	57
No arrangements	2	1
Workplaces with ten or more employees		

Source: Kersley et al. (2006: 204, Table 7.12)

union safety representatives. Such consultation could be either through elected representatives or directly with individual employees.

The effect of this new legislation became clear from a later WERS survey. Using a new categorisation of health and safety arrangements, the 2004 WERS indicated that since 1998 there had been a shift from joint committees dealing with health and safety and an increase in resort to so-called direct methods. In fact there had been a drop in the established means of giving employees formal voice – from 51 to 42 per cent of workplaces – and a rise in so-called direct methods from 47 to 57 per cent (Table 1.1).

In their presentation of these data Kersley et al. (2006) were careful not to refer to 'direct consultation' (the category of consultation brought into existence by the 1996 regulations). Advisably so, because the category they did use ('direct methods') is a ragbag. It includes not only 'consultation directly with the workforce' but management chains, cascades and staff meetings, and also the use of newsletters, notice boards and email. The term 'direct methods' thus contains the possibility that what takes place may not, in any meaningful sense, be consultation at all but just the more or less substantial one-way provision of information from management to employees. Whatever the precise content of 'direct methods', however, they became more widespread between 1998 and 2005 at the expense of joint committees and worker representatives, the use of each of which arrangement for health and safety consultation declined.

The presence of 'direct methods' of consultation on health and safety is a function of workplace size. Such methods are much more common in smaller workplaces; joint committees are much more common in larger workplaces; whereas there is no such clear pattern for employee representatives (Figure 1.2). However, health and safety arrangements are not only a function of size, they are also affected by union recognition.

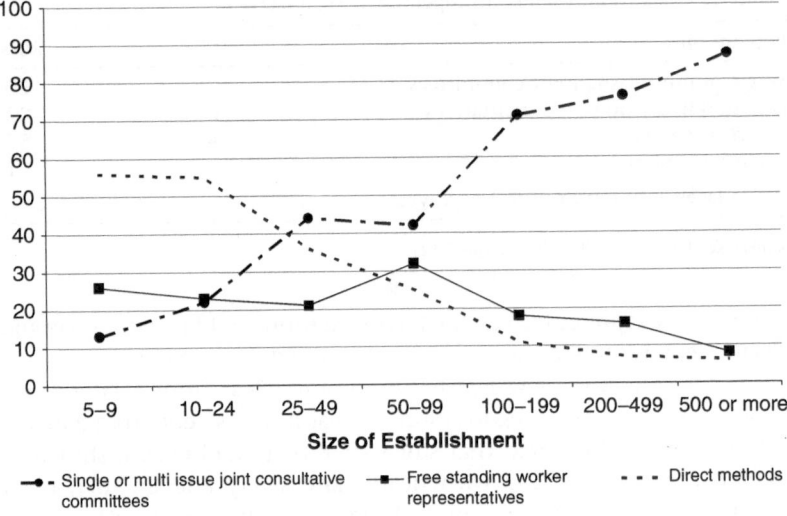

Source: WERS 2004: Management data set

Figure 1.2   Presence of HSCs

Table 1.2   Use of 'Direct Methods', size of establishment and union recognition

Percentage	Size of establishment (employees)						
	5–9	10–24	25–49	50–99	100–199	200–499	500 or more
No recognised trade union	65	77	49	38	27	26	10
Recognised trade union	56	55	36	25	11	7	6

Source: WERS (2004 Management data set)

On average, workplaces which lack union recognition are consistently more likely to resort to so-called direct methods, even within the same size bands (Table 1.2). And of course trade union recognition has been falling since the 1980s. In 1980 64 per cent of establishments with 25 or more employees had recognised trade unions. This share had fallen to 42 per cent by 1998 and to 39 per cent by 2004, with considerably lower figures for the private sector than for the public sector, for smaller establishments and for particular industries (Millward et al., 2000: 96, Table 4.5; Kersley et al., 2006: 120, Table 5.3).

For a considerable time it has been apparent that, other things being equal, three factors made for a *structure of vulnerability* at least as far as industrial injuries in manufacturing are concerned – small size of workplace, absence of trade unions and management alone dealing with health and safety – and that the opposite combination – large size, unionism and worker representation – made for a beneficent constellation (Nichols, 1986; also Nichols, 1997: Part 2). Recently, we conducted a statistical re-analysis of 1990 WIRS data, which is directly relevant to these issues (Nichols et al., 2007; Walters and Nichols, 2007: 30–40). This sought to improve technically on previous multiple regression analyses.[1] It suggested with a fair degree of robustness that, as judged by manufacturing injury rates, it is significantly better to have health and safety committees with at least some members selected by trade unions than to have such committees with no members selected by trade unions, which suggests that there is a mediated trade union effect on safety; and that the presence of health and safety representatives also has a beneficial effect. In short, this suggests that employee participation *matters* – and this after controls had been made for a number of variables, namely, the percentages of manual and female employees, industry and region, union density and also size of establishment (which, as in many other studies, was found to have a negative relation to injury rate, small not being beautiful in this connection).

Whether we consider the falls in size of establishment, the decline in trade unionism, or, specifically, the partly related changes that have taken place in relation to the consultation arrangements for employees, these processes appear, other things being equal, to make employees more vulnerable to injury and ill health at work. Some of these processes have been driven by the desire of governments to keep trade unions in a weakened state and to increase employer flexibility or decrease regulation, and are a function of the neo-liberal leanings and commitments of British political parties since the 1980s. But whereas the consequences of such processes – a lack of collective support for employees and an increase in individualised and informal consultation practices – are to be regretted, there is no reason to assume that all is well with the preferred model itself.

The preferred model in practice

Multiple regression models such as the one referred to above do of course necessarily stand at some distance from particular workplace level processes and practices and from the particular environments in which they take place. It was for this reason the greater part of our book *Worker Representation and Workplace Health and Safety* (Walters and Nichols, 2007) was taken up with reporting case studies that we conducted between 2001 and 2003, some of which we will draw upon shortly.

Altogether we studied 15 workplaces in Chemicals, Construction and Retail. Some of the results can be readily communicated. For example:

- There was a positive relationship, in each industry and overall, between employee ratings of how well managements kept people up to date on general employment matters, provided them with a chance to comment, responded to suggestions etc and how well they did so on health and safety specifically.
- Employment status mattered. In Construction, for example, manual workers employed by the main employer were generally more likely to be consulted on various health and safety matters than were those employed by subcontractors on the same site or agency workers.
- Despite the emphasis recently placed on the economic case for health and safety – the provision by the Health and Safety Executive (HSE) of a health and safety ready reckoner for employers, for example – in none of the case studies was there an arrangement in place to cost injuries or ill health or their prevention.

It is not the above general claims that concern us here, however, but the fact that the case studies in one of the industries, Chemicals, constitute a critical case for the examination of the preferred model in practice. As with other industries, in Chemicals we chose five establishments in which we were able to gain good access to management, personnel and documentation. We especially selected establishments in which there were thought to be reasonably good records of injuries in order that we could use them as one measure of health and safety performance.

The research in each case study comprised:

- documentation provided by the companies and trade unions involved;
- interviews with a range of personnel in each establishment including senior managers, health and safety managers and advisers, supervisors, safety representatives, shop stewards, manual and non-manual workers; and
- a questionnaire-based survey of employees in each establishment. In Chemicals overall 1,477 employees were surveyed with response rates varying between 40 and 80 per cent.

There were two large establishments (over 700 and about 350 employees), one medium-sized establishment (about 270) and two small establishments (just over 100 and just below 50) These were engaged in a range of activities involved in the manufacture and supply of chemical products. One of the two large establishments manufactured pharmaceuticals. All the establishments were part of larger companies that had more than one UK establishment. All five establishments recognised unions, which had above-average densities for the industry. In all except one case there were formal structures for collective bargaining and consultation.

The model of representation and consultation found under the SRSC Regulations applied in each of the five establishments. Consequently, every one of them should have consulted safety representatives in good time with regard to:

- the introduction of any measure at the workplace which may substantially affect the health and safety of the employees represented by the safety representative concerned;
- arrangements for appointing or, as the case may be, nominating competent persons to advise on health and safety matters;
- any health and safety information required to be provided to employees;
- planning and organising health and safety training required for employees; and
- the health and safety consequences of the planning and introduction of new technologies into the workplace.

As already indicated, by the preferred model we mean the case framed by the SRSC Regulations of 1977. Under these, where there are recognised trade unions the employer is required to make arrangements to consult over health and safety with the representatives that such unions have appointed. In the residual legal case, as it might be considered, in which no unions are recognised, employers are required to make arrangements to consult, either directly or through representatives that the workers have elected under the 1996 Regulations. In short, in these case studies we should have found representation and consultation on the implementation and operation of a range of health and safety management arrangements that took the form of two-way communication, conducted in good time and which implied that action would be subsequently taken. This was not the case in the majority of situations we found in our case studies.

In some cases, in the biggest workplace and in a medium-sized one, there were clearly representational and consultative practices on health and safety issues that were working to the satisfaction of the health and safety representatives and the workers they represented. These were the same case studies where there was also strong evidence of a conspicuous commitment to such approaches on the part of *senior* management; and in some cases it was also clear that trade union health and safety representatives in these workplaces would have been unable to function effectively in the absence of management commitment to participation.

Without such commitment, factors that promoted the operation of representative participation either did not exist or had a limited operational capacity. Such factors included, for example:

- properly constituted joint health and safety committees at site and departmental level;
- accountability of managers to the joint health and safety committee;
- engagement of health and safety representatives with the health and safety practitioners from the safety health and environment departments;
- dialogue with local area and line managers within the establishment and health and safety representatives;
- the provision of facility time to undertake health and safety representative functions such as joint health and safety inspections, investigations of workers complaints, making representations to managers and so on;
- involvement of health and safety representatives in risk assessment;
- involvement of health and safety representatives in reporting and monitoring occupational health and safety;
- access of health and safety representatives to workers; and
- access to training for health and safety representatives.

In the case studies where management commitment to participatory approaches was poorly developed, such as in the case study with over 100 employees and to some extent in the case studies with fewer than 50 and with 350 employees, these kinds of arrangements were either absent or set up in very limited ways.

Three things stood out.

First, there was limited development of the consultative structures and processes themselves. For example, in the workplace with just over

100 employees, which produced liquid toiletries, there was a requirement that health and safety should feature regularly at board level within the organisation – but it had a poorly developed health and safety management system. Senior managers at the establishment regarded health and safety management as being about 'following procedures'. They had little understanding of the issues involved, having received little in the way of training on health and safety. They relied almost entirely on the health and safety manager – who had been appointed following an HSE enforcement action – to take responsibility for organising health and safety at the establishment. Although there was trade union representation at the company and a health and safety representative had been appointed, the management did nothing to encourage or support worker representation or consultation on health and safety matters. The safety committee for the establishment was not constituted or operated according to the guidance that accompanies the SRSC Regulations; and its very existence appeared to be unknown to a substantial proportion of the workers who responded to the questionnaire. Trade union representatives reported:

> Our views are ignored by the health and safety manager and by management generally.

> There is little communication or involvement with management.

> There is no consultation.

> Management don't consult over health and safety issues.

Second, there was little in the way of training on health and safety for workers and little evidence of meaningful direct consultation at the establishment, or even of provision of information on health and safety matters.

Another notable feature of these establishments was the inability of health and safety representatives to find time to engage fully with these structures and processes, or to receive training to do so. Both aspects were under the control of management and dependent on its will and capacity to facilitate such participation.

Third, although most establishments were covered by the Control of Major Accident Hazards (COMAH) Regulations and subject to greater than average scrutiny by the HSE, there was no evidence of the influence of the regulatory agency intervening in matters covered by the SRSC Regulations that could be anticipated to involve safety representatives. Implementation and operation of the regulations were therefore more dependent on the wider relationship between the trade unions and the management than any external enforcement pressure.

Despite the applicability of the SRSC Regulations to the workplaces represented by our case studies, certain requirements were repeatedly under-implemented. Even in the workplace where arrangements were the best developed, they still fell short of what is provided for in the regulations in a number of important respects. These included, for example, consultation over the appointment of competent persons, over training and over the introduction of new technologies. In the other workplaces, health and safety representatives' experiences of the operation of the legislative requirements were even more limited. They ranged from experiencing difficulties in obtaining information, time and facilities to undertake practically all aspects of their functions to not taking part in more specific activities such as risk assessment or joint inspections. Substantial concerns were also expressed in these establishments over the extent to which the views of representatives were acted upon by managers.

In nearly all the case studies where the SRSC Regulations applied, the activities of health and safety representatives fell short of their potential in many areas defined in the legislation.

The legal model on which the development of these regulations was based adopted a set of assumptions about the capacity of their beneficiaries to ensure their application without the further intervention of either the law or the regulatory agencies.

Apart from anything else, then, there is a case for increasing the role of regulatory agencies in ensuring compliance with the legal requirements for representation and consultation in those cases where the SRSC Regulations apply. Yet under the present legal framework, as already pointed out, these Regulations only apply in the limited proportion of establishments in which trade unions are recognised. The great majority of employees in the private sector are not employed in workplaces where trade unions are recognised, and there is most certainly a case for improving the consultation and representation in them. The Health and Safety (Consultation with Employees) Regulations make a very weak contribution to this, not only because they confer lesser rights on representatives of employee safety than the SRSC Regulations but because they permit so-called direct consultation. In this form of consultation the employee faces the employer on a one-to-one basis. The collective leverage (and quite probably the greater access to information and awareness that comes with collective forms of representation) is thereby lost. The individual faces the employer (who commands a collective organisation) alone.

The transposition into the UK of another European Directive, the Information and Consultation Directive 2002, might have marked at least some advance had it linked systems of health and safety representation to broader mechanisms of representation (James and Walters, 2005: 122). But the UK Information and Consultation Regulations 2004 (the so-called ICE Regulations) do nothing to integrate health and safety employee representation into any such wider provision. In fact, they give employees the right to be informed (but not consulted) about an undertaking's economic situation; to be informed and consulted about employment prospects; and to be informed and consulted with a view to reaching decisions likely to lead to substantial changes in work organisation or contractual relations, including redundancies and business transfers. They say nothing of health and safety.

The ICE Regulations were conceived in their UK transposition within a problematic that stressed the gains and profit from reduced labour turnover and increased productivity. The regulatory impact assessment suggested benefits net of costs in the region of 'hundreds of millions of pounds over ten years' and pointed approvingly to the cost and regulatory impact being very considerably lessened because the regulations excluded enterprises that employed fewer than 50 employees, these amounting to almost 97 per cent of the enterprises in the UK (DTI, 2004: paras 24, 43–55). By April 2008 the Regulations were expected to have covered about three quarters of employees (rather than workplaces) in the UK, but they were deliberately conceived to facilitate diversity of provision, including no provision at all (there being no obligation for an employer to initiate an information and consultation procedure unless a request has been made by at least ten per cent of the workforce). Not only is it the case that employee rights to information and consultation under this legislation do not apply automatically, with ten per cent of an undertaking's employees being required to trigger negotiations on an Information and Consultation Agreement, but this requirement is subject to a minimum of 15 and a maximum of 2,500 employees, which means that, in the case of a workforce of only 50 employees, the real threshold may be as high as 30 per cent. On top of this, and to weaken rights even further, the employer may regard part-time employees as half rather than whole persons. In short, as convincingly shown by Ewing and Truter (2005), who write aptly of 'voluntarism's bitter legacy', these Regulations are highly complex since there are several different options for drawing up new consultation arrangements (including, as noted, not drawing up any arrangements), they are not always clear and they are hedged about with qualifications to the detriment of employees.

The health and safety-specific HSCE and SRSC Regulations are not as complex as the ICE Regulations; but a survey of employers provides little reason to be sanguine about their knowledge of them. It reports that only a minority of employers were aware of either the HSCE or the SRSC Regulations; that there was 'little evidence to suggest that the regulations, particularly the HSCE regulations had a major impact on workplace practice' and that 'most employers and employee representatives felt that the Regulations have made little difference' (Hillage et al., 2000: 83).

In the UK, therefore, there are problems with the preferred model. Beyond this, there are yet greater problems with the legal consultation and information requirements for those in workplaces that lack an effective trade union presence.

Notes

1. Briefly, as compared with Reilly et al. (1995), we reduced the large number of regional and industry dummies to make a more robust model; reduced the number of independent variables, some of which rested on fine and unclear distinctions; used a Poisson count method instead of a Cox zero corrected method (which entailed adding a bit to the many zero observations); and tested for endogeneity and interaction effects.

2
The Australian Framework for Worker Participation in Occupational Health and Safety

Richard Johnstone

This chapter examines the provisions governing worker participation in occupational health and safety (OHS) in Australia. Australian OHS regulation has been significantly influenced by the UK Robens model, though in some aspects some of the Australian OHS statutes have gone further than Robens and have adopted Scandinavian approaches to OHS regulation – most particularly in relation to the powers of health and safety representatives. On the other hand, although some of the Australian OHS statutes gave a central place to trade unions in the implementation of the health and safety representative provisions, since the early 1990s these provisions have been gradually removed, so that none of the Australian OHS statutes now gives trade unions a privileged role in worker participation in OHS. The chapter also discusses another approach to worker participation in Australian OHS law – statutory rights of entry vested in trade union officials to investigate (and in one state to prosecute) contraventions of OHS statutes duties by employers.

A discussion of the Australian provisions is complicated by the fact that Australia is a federation, currently with six states (New South Wales, Victoria, Queensland, South Australia, Western Australia and Tasmania) and two internal territories (Northern Territory and the Australian Capital Territory, ACT). The Australian Constitution provides the federal government with no express power to legislate for OHS,[1] although Australia's ratification in 2004 of ILO Convention No 155[2] would enable a federal government to legislate for OHS using the 'external affairs' power (Section xxix of the Constitution), and recent decisions of the High Court[3] have considerably broadened the 'corporations power' in the Australian Constitution (Section xx) so that the Federal government could use that power to enact a single OHS statute covering OHS in all corporations. At present, the federal government has not explored these expanded powers in relation to OHS regulation, so that OHS, and workers' compensation, regulation is dealt with at a state or territory level, apart from federal statutes covering federal employees and employees in the maritime industry.

In March 2008 the Council of Australian Governments (encompassing the federal, state and territory governments) agreed that OHS harmonisation was a top priority; that this commitment was to be reflected in an intergovernmental agreement (which was concluded in July 2008); and endorsed a national OHS review (which was initiated in April 2008) to develop a model OHS Act to harmonise OHS legislation within five years. The intergovernmental agreement requires the federal, state and territory governments to enact a model OHS Act which will be drafted by Safe Work Australia based on the proposals of the National OHS Review Panel, and must be agreed to by the Workplace Relations Ministers Council. At the time of writing, the National OHS Review panel had not yet reported on its proposals for worker consultation, representation and participation.

Until OHS statutes in Australia are harmonised through this process, there are ten major Australian OHS statutes, one for each state and territory, and two federal statutes, one for employees of federal government departments and Commonwealth authorities, and the other covering the maritime industry.[4] There are also specialist OHS statutes covering the mining industry in some states. This chapter examines provisions for worker participation in OHS in the nine general OHS statutes, and places greatest emphasis on the provisions in the more populous eastern states: New South Wales, Queensland and Victoria.

Workplace participation provisions in the Australian OHS statutes

Each of the Australian OHS statutes imposes a duty upon employers to consult with employees in relation to OHS, and each establishes institutions for worker participation. At workplace level the primary participatory mechanisms provided for under the Australian OHS statutes are employee health and safety representatives (HSRs), which are to be found in all jurisdictions (apart from the 1986 Northern Territory Act, which ceased to have force in mid-2008), and workplace health and safety committees (HSCs).

Broadly speaking there four models of employee representation in the Australian OHS statutes: the old Northern Territory model, which makes provision for OHS Committees only (and which was superseded in mid-2008); the New South Wales (and new Northern Territory) approach, which imposes a general duty on the employer to consult employees, but offers great flexibility in relation to the participatory institutions which might be adopted; and the model adopted by the other jurisdictions, which makes detailed provision for the election, functions and powers of HSRs and for OHS committees. Of these Queensland vests HSRs only with consultation, negotiation and information rights (the third model), while the Commonwealth (Cth), Victoria, South Australia, the Australian Capital Territory (ACT), and to a lesser extent WA, Tasmania and now the Northern Territory provide HSRs with enforcement powers (the fourth model), principally powers to issue 'provisional' improvement notices and, in the case of the Commonwealth, Victoria, South Australia and the ACT, powers to direct that dangerous work cease.

(a) The Northern Territory

The *Work Health Act 1986*(NT) (section 29(4(d)) required an employer, so far as is reasonably practicable, to consult relevant workers about the development of measures to promote OHS. The Act, however, made provision only for OHS committees (section 44A), which had to be established by the employer who employed more than 20 employees if so requested by a majority of workers at the workplace. A committee comprised both members elected by workers (worker members must constitute at least half the members of the committee) and members appointed by the employer. The functions of the committee (section 44C) included facilitating consultation and cooperation between the employer and the workers in initiating, developing and implementing measures designed to ensure the health and safety of workers; keeping itself informed about OHS standards and reviewing and making recommendations on OHS rules and procedures at the workplace; recommending to the employer the establishment, maintenance and monitoring of OHS programmes, measures and procedures at the workplace; and considering and making recommendations in relation to changes at the workplace that may affect OHS. A person nominated by the committee could inspect the workplace or part of it at times agreed with the employer (section 44D).

The 1986 Act contained no provisions for HSRs, although a review of the Act in 2007 (Shaw et al., 2007) recommended that provisions for HSRs be introduced. In mid-2008 the *Workplace Health and Safety Act* 2007 (NT) came into force. The worker participation provisions in the new statute (see Part 4) are similar to those in the New South Wales Act discussed in the next section, although the new Northern Territory provisions also provide HSRs with powers to issue a 'notice of safety hazard' and to direct that work cease if it poses a 'serious and immediate risk to a worker's health and safety'.

(b) New South Wales

Until 2000 the New South Wales OHS statute also confined worker participation on OHS to health and safety committees (HSCs), although committee members were given some inspection and OHS information powers (see Brooks 1993, pp. 447–62). The approach taken in the *Occupational Health and Safety Act* 2000 (NSW) (OHSA (NSW)) (sections 13–18, which are fleshed out by clauses 21 to 26 of the *Occupational Health and Safety Regulation* 2001) is to give the employer and its employees flexibility in their choice of workplace arrangements. As the following description illustrates, the New South Wales approach to worker participation in OHS deviates quite significantly from both the classical Robens model and the provisions in the other Australian states.

The OHSA (NSW) imposes a general duty on the employer to consult the employer's employees to enable employees to contribute to the making of decisions affecting their OHS. 'Consultation' is defined as the sharing of relevant OHS information; giving employees opportunities to express their views and to contribute to the resolution of OHS issues; and the employer valuing and taking into account these views. Consultation is required when risks are assessed, control measures determined and monitoring measures devised; when decisions are made about the adequacy of facilities for the welfare of employees; where changes are made that might affect the OHS of workers; and where decisions are made about procedures for consultation.

The Act allows significant choice in the ways that consultation may be undertaken. First, consultation may take place with an OHS committee at the employer's undertaking or workplace. A committee must be established if an employer employs more than 20 employees and a majority of employees request the formation of committee, or the regulator requests or the employer initiates the formation of the committee. Second, consultation may be undertaken with HSR(s) elected by employees. HSRs must be elected if one employee so requests or the regulator directs. Employers may initiate processes to elect HSRs. The third option is for participation to take place in accordance with other arrangements agreed to by the employer and employees.

A federal or state trade union can represent employees under the agreed arrangements.

An employer is to consult employees on the type of representation to be used at the workplace; and the regulations set out processes to be followed in establishing such arrangements (for example, how work groups represented by HSRs are to be determined, the election process and so on), in implementing them (how representation will take place) and in periodically reviewing them. The minimum requirements for the election of HSRs are similar to other Australian provisions (see below), but regulation 23(4) provides that where a work group is represented by both an HSR and an HSC, the HSC 'is the principal mechanism for consultation for that work group'. This is a notable departure from the classical Robens model, which accords primacy to HSRs. Further, unlike the Robens model as implemented in the other Australian states (see below), in New South Wales HSRs and HSCs have exactly the same limited functions (principally to review OHS measures and to investigate and resolve matters) and are owed similar duties by an employer. HSRs and HSC members must undergo OHS training.

(c) Worker representation and participation in the other Australian OHS statutes

The Commonwealth, Victorian, Queensland, South Australian, Western Australian, Tasmanian and ACT OHS statutes generally prescribe a two-stage approach to the election of HSRs. In the first stage, employees can negotiate with their employer as to how many HSRs there will be, and their areas of representation (often referred to as the HSR's work group). The idea of specifying an HSR's area of representation emerged in the early 1980s as a response to employer concerns that, where many trade unions had coverage of a workplace, each would want to have an HSR at the workplace. Negotiating work groups is a way of clarifying the 'jurisdiction' of each HSR. The second stage involves the process for electing HSRs for each work group. Once elected, HSRs are vested with inspection, consultation and information powers – and in some of the statutes (discussed in the next section) HSRs are given powers to enforce the OHS statute. The provisions differ in detail from statute to statute, and there are differences in the way that the process is initiated, in the extent to which and manner in which the statutes draw trade unions into the process, and in whether the processes are confined to the 'employees' of the 'employer'.

An example of provisions governing the election of HSRs are sections 70 to 85 of the *Workplace Health and Safety Act* (Qld) (WHSA (Qld)).

The employer must tell new workers, and display notices, about the provisions in the Act dealing with workplace HSRs and committees. Workers at a workplace (represented by a trade union if they so choose and so inform the employer) may negotiate with their employer about HSRs, including the number of HSRs, their area of representation, the frequency of inspection, access to training and similar matters. The employer *must* negotiate if so asked, and may not exclude from any negotiations any union that has workers at the workplace, if the workers have told the employer of their preference to be represented by the union. A tribunal, namely, the Queensland Industrial Relations Commission, may hear and decide an application by any person aggrieved by the failure of such negotiations.

HSRs are elected by workers, and cannot be appointed by the employer. Workers at a workplace are entitled to elect one or (as a result of negotiating with their employer) more HSRs on their own initiative or at their employer's suggestion, and must tell their employer of their intention to conduct the election, and of the result as soon as practicable after they have elected representatives. The workers can ask the employer to facilitate the election or can ask any union represented at the workplace to conduct the election for all workers. A worker does not need to have any particular qualification to become a HSR. A worker is elected for a term of two years, and may be re-elected. The employer must display the identity of elected HSRs within five days of the election. The HSR ceases to be an HSR if she or he resigns or ceases to be a worker at the workplace.

The Queensland provisions are fairly typical of those in the Australian statutes, but there are differences in approach that are worth noting. The Queensland provisions refer to 'workers' (which I will define and discuss further later in the chapter), while all of the other provisions refer to 'employees' – that is, persons employed by an employer under a contract of employment. In the *Occupational Health and Safety Act 1989* (ACT) (OHSA (ACT)) employers in workplaces with ten or more employees *must* begin processes of negotiating designated work groups, and electing HSRs – in all other statutes, employee(s) initiate the process (but in Tasmania only in workplaces with more than ten employees) and in some statutes, most notably the *Occupational Health and Safety Act 2004* (Vic) (OHSA (Vic)) and the *Occupational Safety and Health Act 1984* (WA) (OSHA (WA)), the employer can initiate the process. Once the process is so triggered the OHS statutes enable employees to negotiate with the employer to form work groups, taking into account a range of factors (including the number of employees, type of work, nature of risks, overtime or shift work arrangements, and so forth), and most enable the regulator or a tribunal to resolve contentious issues.

As illustrated by the Queensland provisions set out above, each statute outlines a process for employees in each work group to elect one (in Victoria there can be more than one) of the employees in that group as an HSR for the group. In the OSHA (WA) and the *Occupational Health and Safety Act 1991* (Cth) (OHSA (Cth)) the parties can agree for the election to be conducted by an electoral commissioner appointed under the relevant electoral legislation; and in the OSHA (WA) a trade union can conduct the election. Most of the OHS statutes provide for the election of deputy HSRs. All of the OHS statutes specify the period for which HSRs are elected, and for the disqualification of HSRs for abuse of powers causing harm to employers. The *Occupational Health, Safety and Welfare Act 1986* (SA) (OHSWA (SA)) provides that an HSR can be disqualified for negligent underuse of powers.

One notable aspect of this brief description of the processes for the election of HSRs is that there is very little mention of trade unions. The original Commonwealth, Victorian and Western Australian statutes did establish a 'two-track' model, in which trade unions did the negotiating for the formation of work groups, and oversaw the election process. It was only if there were no unions with coverage of the workplace that employees conducted the processes. Each of the two-track approaches has been dismantled by conservative governments, starting with Victoria in 1993 and ending with the Commonwealth in 2004. Each of the statutes now has a single process, although, as we noted above, the OSHA (WA) does enable a trade union to conduct the election for HSRs. Some of the OHS statutes require trade unions with coverage of the workplace to be consulted, or, as in the cases of the OHSA (Cth), WHSA (Qld) or OHSWA (SA), enable employees to ask a trade union to negotiate on their behalf.

Once elected (or selected in the OHSA (Cth)), HSRs have a wide range of powers generally to be exercised in relation to the work group only, and with correlative duties imposed upon employers. Most of the powers and functions of Australian HSRs are similar to those vested in HSRs under the *Health and Safety at Work Act 1974*, and include:

- the right to inspect the workplace;
- the right to be consulted where workplace changes affect OHS;
- the right to be present, with the consent of the employee, at interviews between an employee and employer/inspector;

- the right to accompany an inspector on an inspection (or at least to be told of the presence of an inspector at the workplace);
- the right to information affecting the OHS of employees;
- OHS training and facilities;
- time off work to perform HSR functions;
- the right to assistance from OHS experts (but not in the Queensland, Tasmanian, Western Australian or ACT OHS statutes); and
- the right to investigate complaints about OHS-related issues.

Further, HSRs can always ask an inspector to visit (in some statutes this is explicitly included as a HSR power).

Generally the Australian OHS statutes provide that an HSR incurs no liability by virtue of her or his activities as a HSR. The statutes also seek to protect employees from discrimination substantially on the ground that they raised OHS issues or exercised their functions and powers as an HSR.

Health and safety committees

One of the powers of an HSR in the OHS statutes described in the previous section is to request that the employer establish a health and safety committee (HSC). In Western Australia an HSC can also be requested by an employee, and in Tasmania by a majority of employees in a workplace of more than 20 employees. Typically the OHS statutes provide that employer representatives must not exceed the number of employee representatives, and generally (apart from specifying a few fundamental requirements such as the minimum number of meetings each year) leave it to committees to determine their own processes. The functions of HSCs are fairly standard across the OHS statutes, and generally include assisting the employer to develop, implement and review OHS measures; facilitating cooperation between employer and employee in relation to OHS matters; and assisting the employer to disseminate OHS information to employees.

Research into the operation of the HSR and HSC provisions

It is not clear just how many HSRs and HSCs there are in Australian workplaces, as there has not yet been a comprehensive study of HSRs in Australia. The 2004 Maxwell Report in Victoria (2004, p. 195, para. 886) noted that, although the OHS regulator had 'no record of the numbers or distribution of HSRs, it is generally accepted that the majority of

Victorian workplaces do not have HSRs'. In the piecemeal research on Australian workplace participation that has been conducted to date, there have been a range of surveys, including by academics, consultants and union peak bodies. The most recent, and final, Australian Workplace Industrial Relations Survey (Moorehead et al., 1997) found that 66 per cent of workplaces surveyed had elected HSRs, 43 per cent had HSCs in place, and 30 per cent had no HSRs or HSCs. A survey by the Australian Council of Trade Unions (Australian Council of Trade Unions, 2005) found that 81 per cent of HSR respondents reported that their workplace had an HSC (and 14 per cent reported that their workplace did not have a committee), but 19 per cent of those reporting the existence of an HSC said it did not work 'properly and well', and 13 per cent were not sure how the HSC worked. Vanderkruk (2003) found that in a randomly selected group of Queensland workplaces 21 per cent had HSRs; and that the bigger the workplace (and, in particular, if the workplace also had a workplace health and safety officer), it was more likely that there would be HSRs. Vanderkruk also found that, even though the WHSA (Qld) required HSRs to be elected (see above), many in her study had been appointed by the employer, a finding supported by Bos (1995), who found that 58 per cent of HSRs in Queensland had been appointed, not elected. A South Australian study (Blewett, 2001) similarly found that in South Australia 22 per cent of HSRs had been appointed by the employer, rather than elected, even if not interested in the position.

The limited available research data suggests that HSRs do not receive adequate OHS training. For example, only 16 per cent of HSRs surveyed in Vanderkruk's (2003) Queensland study had received training. The ACTU's 2004 survey (Australian Council of Trade Unions, 2005) reported that 30 per cent of respondent HSRs (and a previous (2001) survey suggested 25 per cent) had not received OHS training in past two years, and that five per cent (25 per cent in 2001) had received no training at all. Of those receiving training, the Australian Council of Trade Unions (2005) found that 77 per cent of HSR respondents said that their training was useful in helping them to carry out OHS functions.

The HSR surveys suggest that, in seeking to carry out their functions, HSRs met with some resistance from employers, including bullying and harassment. The Australian Council of Trade Unions (2005) survey revealed that 55 per cent of surveyed HSRs believed that they did have enough time to perform their HSR functions, while 33 per cent believed that they did not have enough time. Nine per cent indicated that they had over ten hours per week to devote to HSR functions, while five per cent reported that they spent from five to ten hours a week on their HSR role, 29 per cent put the figure at one to five hours a week, and 48 per cent reported that it was less than one hour per week. The 2003 Victorian Trades Hall (VTHC) survey of HSRs found that 40 per cent of surveyed HSRs did not consider that they had been provided with access to appropriate equipment, and 50 per cent reported that they had not been provided with copies of the relevant OHS statute, regulations or codes of practice (VTHC, 2004). The ACTU and VTHC surveys found that HSRs reported experiencing significant negative pressure from employers. For example, the ACTU surveys found that 28 per cent of HSR respondents in 2004 (and 20 per cent in 2001) said that they had been pressured by management to not raise OHS issues. The VTHC surveys (2004 and 2006) each reported that about 33 per cent of surveyed HSRs had made similar claims. Bos's (1995) Queensland study found that 60 per cent of HSRs reported that they had encountered some opposition to improving OHS in workplace, and Blewett's (2001) South Australian study found that 49 per cent of HSRs reported difficulty in performing their representation work. Both the 2004 ACTU survey and the 2005 VTHC survey found that 25 per cent of HSRs claimed to have been bullied or intimidated by management because they raised OHS issues; and Blewett's (2001) study reported that 43 per cent of South Australian HSRs had claimed that they had been harassed or bullied because they were HSRs.

More specifically, the HSR surveys suggest that HSRs have significant difficulties in exercising their core functions. For example, the 2004 ACTU survey found that only 41 per cent of HSRs said that their employer automatically consulted them on OHS issues, that 40 per cent of them were consulted only when they asked, and that 13 per cent were never consulted. Both the 2003 and 2005 VTHC surveys reported that 65 per cent of HSR respondents claimed that employers did not automatically consult HSRs. Bos (1995) found that 37 per cent of Queensland HSRs were never consulted. Similarly, the 2004 ACTU survey found that 48 per cent of HSR respondents were automatically informed by an employer when an injury occurred, 35 per cent were informed when they asked and 12 per cent were never informed. Further, only 44 per cent of HSRs in the Australian Council of Trade Unions (2005) survey (37 per cent in the 2005 VTHC survey) reported being automatically included in investigations of OHS incident; 34 per cent reported being included when they so requested; and 17 per cent were never included (16 per cent in the 2005 VTHC survey). The Australian Council of Trade Unions (2005) survey also concluded that 58 per cent of HSRs said that HSRs or other workers were regularly included in employer OHS inspections of the workplace, 31 per cent were occasionally included and 8 per cent were never included. Those that were included said that 79 per cent of time the inspections resolved OHS problems. Fifty per cent of HSRs reported that they frequently, and 37 per cent occasionally, carried out inspections of their designated work groups, and 6 per cent never conducted inspections (Australian Council of Trade Unions, 2005). Of those who did inspections, 79 per cent said OHS issues identified were fixed by employers. Bos (1995) suggested that 41 per cent of Queensland HSRs never undertook inspections.

Finally, it appears that OHS inspectors across the jurisdictions tend not to consult with HSRs during workplace visits. For example, the Australian Council of Trade Unions (2002) survey in 2001 found that 11 per cent of surveyed HSRs said that inspector automatically spoke to them when the inspector visited their workplace; most said that an inspector spoke to them only when HSR initiated the contact. My own research, with Michael Quinlan, into the OHS inspectorates in Queensland, Victoria, Western Australia and Tasmanian involved field visits with inspectors; and our observations were that inspectors spoke to HSRs in only 37 per cent of workplace visits. Against that, Sweeney Research (2005) reported that 66 per cent of HSRs surveyed by telephone said that inspectors provided satisfactory support to HSRs.

Vesting HSRs with enforcement powers

As I have already noted, the rights and powers of HSRs in New South Wales and Queensland are limited to OHS information, inspection of the workplace and consultation on OHS issues. These statutes share the Robens vision of employers and workers largely having a 'great natural identity of interest ... in relation to safety' (Robens, 1972, para. 66) so that the state's approach to developing an 'a more effective self-regulating system' involving workers (Robens, 1972, para. 41) has been limited to 'constructive discussion, joint inspection and participation in working out solutions' (Robens, 1972, para. 66). Indeed, the Robens Report suggested that if an HSR discovered an OHS issue during an inspection 'we do not believe that any responsible employer would ignore a genuine problem revealed by such inspection' (Robens, 1972, para. 65). Not everyone would agree with this assumption, and some would argue that managers and workers will not always agree on what should be done to resolve an OHS issue, because of the 'inherent and inevitable conflict between health and safety and other production objectives (Gunningham, 1985, pp. 47–8). Certainly, this view was taken by a number of Australian Labor Party governments at federal, state and territory levels in the period 1985 to 1991, so that in the Australian OHS statutes, apart from those in New South Wales, the Northern Territory and Queensland, HSRs have also been vested with important enforcement powers: provisional improvement notices and the right to direct that dangerous work cease. Although the precise nature of these enforcement powers varies from statute to statute, these powers can be grouped together as key elements of a fourth model of worker participation, in which 'self-regulation' as envisaged by Robens has shifted towards an 'enforced self-regulation' model, where enforcement includes provisional improvement notices (PINs) and work cessation directions by HSRs, with the OHS inspectorate playing both an adjudicatory role (reviewing decisions to issue PINs or to stop dangerous work) and an enforcement role (ensuring notices and work cessation orders are complied with).

For example, where an HSR reasonably believes that there has been a contravention of the OHS Act or regulations, the OHS statutes of the Commonwealth, Victoria, South Australia, Western Australia, Tasmania and ACT give the HSR the power to issue a provisional improvement notices (called 'default notices' in South Australia and 'written directions' in Tasmania), requiring the contravention to be rectified within a specified period. HSRs usually can issue such notices only after consultation with the employer to rectify the contravention. The person issued with the notice can have the notice reviewed by an inspector, and the inspector can confirm, modify or disallow the notice. If a PIN is not challenged, or is challenged and confirmed or modified, it is an offence to fail to comply with the provisions of the PIN. The new Northern Territory Act, however, takes a different approach. It vests HSRs with the power to issue a 'notice of safety hazard', but it is not an offence not to comply with the notice – rather, an inspector must investigate and take any necessary action to eliminate or mitigate the hazard.

Despite much employer concern that PINs may be abused, there is little evidence that this is the case. The Australian Council of Trade Unions (2004) survey of HSRs reported that 11 per cent of HSRs had said that they had issued a PIN or default notice, and that 91 per cent of these said it was effective in resolving the OHS issue. Given that HSRs in New South Wales and Queensland would not have the power to issue a PIN, one would expect the usage of PINs to be more extensive than the ACTU survey suggested. Consistent with this, the VTHC (2004 and 2006) survey found that 25 per cent of respondent HSRs in Victoria had issued PINs, and the 2003 survey found that, of the 25 per cent of HSRs issuing PINs, 45 per cent had only issued one PIN.

Another criticism of PINs has been that, in fact, far from being over-utilised, they are not effectively enforced by the inspectorate. A major difficulty with enforcement is that the inspectorate will have no knowledge that a HSR has issued a PIN unless the employer appeals against the PIN, or the HSR informs the inspectorate that a PIN has been issued, or that it has not been complied with. There was no report of a successful prosecution for a contravention of a PIN until 2007, when the Victorian Department of Education was prosecuted for failing to comply with a PIN issued by an HSR at the Altona Primary School because there were inadequate staff facilities at the school. The Victorian OHS inspectorate had subsequently also issued an improvement notice on the Department for failing to remedy the contraventions identified in the PIN. The Department entered a guilty plea, and was fined A$8,000 for failing to comply with the PIN and with the improvement notice.

The OHS statutes of the Commonwealth, Victoria, South Australia and ACT do not just empower HSRs with the right to issue a PIN for a contravention of the OHS legislation, but go further to give HSRs the power to direct that work causing an 'immediate risk' to the OHS of any person cease. This power is generally included as part of a process for HSRs to consult with employers (and in some statutes with the HSC and an OHS inspector) to resolve an OHS issue. If effective consultation to address the hazard is not possible, the HSR can direct that the hazardous work terminate until the hazard is removed or mitigated. The employer who is the subject of such a direction can allocate the affected workers to other tasks, and can summon an inspector to the workplace to resolve the issue.

The new Northern Territory provisions enable an HSR to direct that work cease if it poses a 'serious and immediate risk to a worker's health and safety', but the employer is not obliged to comply with the direction. If the direction is not obeyed, the HSR must report the matter to the inspectorate, which must take 'any action that appears necessary in the circumstances'.

Once again, despite employer suggestions that these work cessation provisions can be, and have been, abused by HSRs, there it little evidence that this is the case (but see Maxwell, 2004, pp. 192–3). The Australian Council of Trade Unions (2004) survey reported that 21 per cent of HRSs said they had directed that unsafe work cease or stopped work for OHS, and 88 per cent said effective in resolving issue.

Coverage of workers who are not 'employees'

In the past two decades, the pattern of working relationships in Australia, as in most other countries, has changed significantly, with a greater incidence of precarious and contingent work (see Johnstone, Quinlan and Walters, 2005). Australian firms increasingly engage contractors, subcontractors and agency labour. The worker participation provisions described above, with a few exceptions that will be discussed below, do not make provision for these kinds of worker to be included in workplace consultation processes, because they couch processes for the election of HSRs in terms of 'employees' and their 'employer', and generally only 'employees' of the employer at the *employer's* workplace are eligible to stand for election as an HSR and to vote for the HSR (Johnstone, Quinlan and Walters, 2005; and Johnstone, 2004, pp. 488–500). Further, as we have seen in the discussion of the ACT, Northern Territory, New South Wales and Tasmanian provisions above, some provisions operate only if there is a threshold number of employees in the workplace, so that, as work is increasingly outsourced, fewer workplaces will have the requisite number of 'employees' to trigger the worker participation provisions.

There are, however, some provisions in the Australian OHS statutes which do enable workers who are not 'employees' in the technical legal sense to participate in the HSR and HSC processes.

For example, as discussed above, the WHSA (Qld) enables 'the workers' at a workplace to elect one or more HSRs for the workplace (sections 67–74). A person is a 'worker' if the person 'does work, other than under a contract for services', for or at the direction of the employer, even if the worker is not paid for work done (section 11). This broad definition of 'worker' has the potential to involve many workers who are not 'employees' of the employer to be included in designated work groups, to stand for election as an HSR and to vote in an election for an HSR. For example, volunteers and labour hire workers are clearly working at the 'direction' of the host 'employer' even if they are not employees. But an independent contractor is excluded from the definition of 'worker' (see section 11), and therefore cannot be represented. Similarly, 'outworkers' engaged to work at home would not be 'at' the workplace with co-workers, and therefore could also not be represented.

The 2007 Northern Territory Act also enables 'workers' to negotiate work groups and elect HSRs. The Act defines a 'worker' even more broadly than does the Queensland Act to include 'any person who workers in the employer's business' as an employee, apprentice, contractor or subcontractor (or their employee), employee of a labour hire company, volunteer or 'in any other capacity'.

Extensive provisions aimed at including independent contractors in the process are to be found in the OHSA (Vic), where section 44(1)(e) provides that in negotiating designated work groups (DWGs) the employer and employees can negotiate whether an HSR is authorised to represent independent contractors 'engaged' by the employer and their employees. It is likely that this provision will be very broadly interpreted to include subcontractors, sub-subcontractors and parties further down the contractual chain (see Creighton and Rozen, 2007, p. 233). In addition, Part 7 Division 2 provides for the negotiation of DWGs covering more than one employer. Section 47 provides that DWGs may be established which comprise employees of an employer in one or more workplaces, and employees of one or more other employers at one or more workplaces. For example, where an employer engages a principal contractor to construct a new plant at the employer's workplace, and the principal contractor both employs its own employees and engages contractors (which in turn employ their own employees and engage subcontractors), all the employees and all the employers can negotiate a DWG for the construction process. Section 52 makes it clear that these multi-employer negotiations for a DWG have no effect on other arrangements for DWGs made by the original employer with the employer's pre-existing employees at the workplace.

Similarly, the OSHA (WA) provides that a scheme to elect HSRs may include provision for the election of one or more HSRs for one or more workplaces, and may provide that a contractor and a person employed by a contractor be treated as employees of the person who engages the contractor (see section 30B).

A different approach is taken in the *Occupational Health and Safety Regulation* 2001 (NSW), where clause 23(2) specifies that the factors to be taken into account in setting up a work group include the pattern of work of employees, including the representation of part-time, seasonal or short-term employees; the geographic location where employees work, including the representation of employees in dispersed locations such as transport work or working from home; and the interaction of the employees with the employees of other employers (including contractors and labour hire). Further, the New South Wales *OHS Consultation Code of Practice* 2001 at 2.4.2 outlines ways a host employer or principal contractor can facilitate consultation in an array of different scenarios where workers are provided by a labour hire firm, contractors or multiple subcontractors. Examples cited include the provision of meeting space, communication assistance (such as telephone and email access for HSRs), inclusion in OHS consultation and reporting systems, and inter-firm consultation about how to create the most effective lines of communication and consultation (see further Johnstone, Quinlan and Walters, 2005).

Finally, under section 39 of the OHSA (ACT) principal contractors in construction can request an Occupational Health and Safety Commissioner to declare that the provisions in the Act relating to the negotiation of work groups and the selection of, and the powers, of HSRs and HSCs apply in relation to the subcontractor's employees. There is a similar provision in relation to HSCs in the WHA (NT) section 44A.

While these provisions all have the potential to enable workers who are not 'employees' of the 'employer' to be involved in HSR structures, anecdotal evidence from OHS regulators and union officials suggests that to date in practice these provisions have not had much impact in drawing these kinds of workers into HSR structures.

Union right of entry for OHS purposes

The HSR and HSC provisions discussed earlier in this chapter essentially provide employees of the employer with OHS information, inspection, consultation (and, in some statutes, enforcement) rights. But, as we have seen, it appears that HSRs are not to be found in the majority of workplaces, which has led to concern among some policy makers and a heightened interest in other approaches to ensure that workers' interests in OHS are protected. In Victoria, for example, the *Maxwell Report* (Maxwell, 2004, para. 991) stated that 'consultation and participation are essential to the achievement of good health and safety outcomes' and that 'widespread lack of representation of workers' health and safety interests represents a major failure' of the 1985 Victorian OHS Act. Maxwell recommended a general duty on employers' to consult with workers on OHS issues (see above), and a consultative structure (discussed above) enabling workers, by themselves or through an HSR, to participate in OHS management (para. 994). Maxwell was concerned that this ideal 'will take quite some time' (para. 995), and 'in the meantime, the Act should include an alternative mechanism for worker representation: the 'conferral of a right of entry on authorised representatives of unions ... for the purposes of investigating any suspected breach of the OHS legislation' (para. 1018). In other words, in the absence of adequate 'internal' provisions for worker representation and participation, the OHS statute should enable union officials to investigate suspected contraventions of the OHS statute.

With the exception of the New South Wales OHS statute (discussed below), until recently the issue of union right of entry to workplaces for the purposes of OHS was not expressly dealt with in the OHS statutes, and the only relevant provisions were those enabling HSRs to seek the assistance of OHS 'experts' or 'consultants' in exercising the HSR's powers. For example, the OHSA (Vic) (section 58(1)(f)) and OHSA(Cth) (sections 28(2)and 70) enable an HSR, in the exercise of her powers, to be assisted by a consultant. The OHSWA (SA) gives an HSR a restricted right to be accompanied by an approved consultant when making a workplace inspection (see sections 32(1)(i) and (2). None of the other Australian OHS statutes enables HSRs to be assisted by a consultant.

The first Australian OHS statute to include union entry powers was the *Occupational Health and Safety Act* 1983 (NSW) (sections 31AG–AP), and these provisions were retained in the OHSA (NSW) in 2000. Sections 76–85 of the OHSA (NSW) give an 'authorised representative' (an officer authorised under Part 7 of Chapter 5 of the *Industrial Relations Act* 1996) of an 'industrial organisation' (that is, a trade union) limited powers of entry 'for the purpose of investigating any suspected contravention of the occupational health and safety legislation'. While the union representative can enter the premises without notice, she or he must notify the occupier of the premises as soon as is reasonably practicable. The authorised representative must produce the authority provided by the Industrial Registrar when required to do so by the occupier of the premises. The powers include making searches and inspections, taking photographs and making video and audio recordings, requiring the occupier of the premises being investigated to provide assistance and facilities reasonably necessary to carry out the investigation, and requiring the production of any documents at the premises that directly affect or deal with the OHS of persons working at those premises, for inspection. In *Jim Pearson Transport* v *Transport Workers' Union* [2007] AIRC 559 the Australian Industrial Relations Commission held that these powers required an employer to answer relevant questions, as well as producing documents. The authorised representative may request assistance from an inspector. The New South Wales Act also empowers the secretary of a trade union registered in NSW to commence proceedings for a contravention (section 106). To date, about a dozen union prosecutions have been conducted under the New South Wales OHS legislation.

In the last few years, similar provisions have been included in the Victorian, Queensland, ACT and the 2007 Northern Territory OHS Acts, although the details of the provisions (method of appointment, procedures upon entry, powers after entry, and so forth) vary from statute to statute. The Victorian, Queensland, and ACT provisions require an authorised representative to have completed approved training, while the Northern Territory Act requires the representative to have appropriate qualifications and experience. These Acts, however, do not give union officials any enforcement powers. Further, the federal *Workplace Relations Act* 1996 (section 756) provides that union officials having a right of entry under a state OHS act can exercise these powers only if they also hold a permit under Part 15 of the *Workplace Relations Act.*

To date there have been some significant cases interpreting these union entry provisions. A recent decision of the Australian Industrial Relations Commission (AIRC) interpreting the union entry provisions in the *Workplace Relations Act* 1996 (Cth) (*Appeal by Australian Municipal, Administrative, Clerical and Services Union* (C2007/3800), AIRC, Full Bench, 8 February 2008) has stated that employers can refuse the union representatives entry to a workplace if the representative cannot provide unambiguous proof of the existence of reasonable grounds for suspecting a breach of the OHS Act. This decision will go some way to reduce the scope of the union entry provisions, because union representatives will most likely need to provide employers with particulars of each suspected contravention.

Another AIRC decision was the first to suspend a union representative's entry permit for nine weeks for 'abuse of system'. In *Australian Building and Construction Commission* [2007] AIRC 717 the tribunal found the representative had contravened the *Workplace Relations Act* (section 758) for persistent failures, upon request, to produce his entry permit. The representative had argued that he did not need to keep showing his permit on repeat visits to the workplace when he was known to the workplace supervisor. Other abuses leading to the suspension included holding a 26-minute stop-work meeting at a workplace to obtain OHS information without employer approval, and, less significantly, refusing to sign the visitor's book and undergo a site induction at another workplace.

Conclusion

This chapter has analysed the Australian framework for worker representation in OHS. It shows that there is no uniform approach to worker participation in OHS, although a majority of OHS statutes (the Commonwealth, Victoria, South Australia, the ACT, and, to some extent, Western Australia and Tasmania) take a broadly consistent approach. A number of characteristics distinguish the Australian provisions. First, they are remarkably prescriptiive and legalistic, setting out processes in extraordinary detail, and requiring considerable legal expertise for their interpretation. Second, the Commonwealth, Victorian, South Australian, the ACT, and, to a lesser extent, the Western Australian and Tasmanian provisions go beyond the Robens model to vest HSRs with significant enforcement powers, principally provisional improvement notices and the right to direct that work cease if it poses an immediate risk to the health and safety of employees. Third, over time, the role of trade unions in the HSR and HSC participatory processes has been reduced, so that only occasionally in the current statutes is there reference to a trade union. On the other hand, and fourth, the provisions enabling authorised union officials to enter workplaces to investigate suspected contraventions of the OHS statute (and, in New South Wales, to initiate prosecutions for detected contraventions) are important powers enabling unions to play a role in protecting workers' OHS in workplaces – and in a sense they enable union representatives to play a role, albeit poorly defined, as a 'roving' safety representative. Finally, it is clear that we know very little of the actual operation of the Australian provisions for worker particpation in OHS, and that there is plenty of scope for empirical research.

Notes

1. For a discussion of the impact of the Australian Constitution on OHS regulation, see Johnstone (2004, pp. 88–99); and Lee and Quinlan (1994).
2. *Convention Concerning Occupational Safety and Health and the Working Environment*, No 155 of 1981.
3. See *New South Wales* v. *The Commonwealth* (The Work Choices Case) (2006) 229 CLR 1; and *Attorney General (Vic)* v. *Andrews* (2007) 230 CLR 369.
4. The statutes are the *Occupational Health and Safety Act* 1991 (Cth), *Occupational Health and Safety Act* 2000 (NSW), *Occupational Health and Safety Act* 2004 (Vic), *Workplace Health and Safety Act* 1995 (Qld), *Occupational Health, Safety and Welfare Act* 1986 (SA), *Occupational Safety and Health Act* 1984 (WA), *Workplace Health and Safety Act* 1995 (Tas), *Work Health Act* 1986 (NT) (replaced in mid-2008 by the *Workplace Health and Safety Act* 2007 (NT)) and *Occupational Health and Safety Act* 1989 (ACT). For an overview of these statutes, see Johnstone (2004, pp. 80–4).

3
Health and Safety Committees in France: An Empirical Analysis

Thomas Coutrot

The number of occupational illnesses recognised each year in France has grown exponentially over the past 20 years, rising from under 5000 in the late 1980s to more than 40,000 in 2006, including 2000 new cases of occupational cancers and close to 30,000 cases of musculo-skeletal disorders. But these figures do not tell the full scale of the public health problem, because occupational diseases are grossly under-reported, as all those concerned with health and safety at work know (Dares, 2007). Health and safety at work is now a central issue of health policies.

Health and safety at work enablers therefore want to improve workplace risk prevention. The public authorities want to boost research, tighten up regulation and make it more effective, get participants working in a more joined-up way. Business is being urged to invest more in prevention. Trade unions want to expand the powers given to employee health and safety representatives, that is, health and safety committees (HSCs) or shop stewards.

Are employee health and safety representatives effective in helping to safeguard workers' health? It is a question that has been addressed in an extensive international scientific literature. Most – but not all – studies find that worker representation does do some good (Walters and Frick, 2000).

Studies that have made favourable findings can be classified into two kinds. On the one hand are those studies that 'suggest that indicators of objective health and safety performance, such as injury rates, are better in situations in which joint arrangements are in place and/or when trade unions are engaged in worker representation in workplaces'. On the other hand, 'other studies point to associations between the presence of representative structures and indicators of a more systematic approach to OHS management' (Walters, 2003, p. 9). The literature includes studies on the United Kingdom, the United States, Australia, Canada and Sweden. While health and safety committees (HSCs) are generally recognised to be key institutions for workplace risk prevention, empirical evidence on their importance is lacking in France. It is this gap that the present chapter sets out to fill.

Conceptual model and empirical data

In France, labour law provides a firm ground for health and safety representation at workplace level. All firms with 50 or more employees in the competitive sector are supposed to install a *Comité d'hygiène, de sécurité et des conditions de travail* (CHSCT). The CHSCT is a subsection of the *Comité d'entreprise* (work council): its members are elected workers' representatives, and the chair is the employer. When one or several unions are present in the establishment, only union representatives can provide candidates at the first ballot of the election. Only in the case of unions' absence, or if more than 50 per cent of workers do not vote at the first ballot, can non-union candidates stand for election; in consequence, safety representatives are usually also union members, with the exception of smaller firms without unions. CHSCT have legal prerogatives: they must be consulted by the employer about health and safety issues, especially in case of major technological or organisational change. In fact, health and safety committees (CHSCTs) are present in only 77 per cent of establishments with 50 or more employees (16 per cent between 20 and 49 employees): in many establishments without any union representatives, there are no candidates for the work council, and the election cannot occur.

According to a theoretical framework we will draw on for this study, mechanisms that underlie the (lack of) effectiveness of employee health and safety representatives ('safety representatives') are of two kinds: cognitive ('worker knowledge') and political ('worker power') (Hall et al., 2006).

On the cognitive aspect, safety representatives can help to bring out the knowledge that employees have – and management is commonly known to lack – of how work really gets done. Prevention professionals' better knowledge of 'real work' is certainly a big plus for prevention policies. Also, discussion and differences aired in HSCs promote – arguably to differing degrees – risk identification and awareness by participants (management, prevention professionals, employee representatives, employees). Finally, the training given to safety representatives and the experiences passed on within trade union networks external to the establishment can improve collective hazard identification and solution-framing abilities.

The cognitive mechanisms are essential because the mere existence of risks does not mean they will be perceived as such, nor that preventive action will result. Even employees exposed to what appears to an outside observer as obvious risks may use denial as a fear-quelling strategy. In 1984, for example, a high percentage (14 per cent) of truck drivers interviewed for the Working Conditions Survey thought they were at no risk of an accident. (This percentage has dropped sharply in the subsequent years, with an increasing public concern about traffic accidents caused by trucks, and stronger training policies for drivers.) In the 2003 SUMER survey (see below) 8 per cent of employees were exposed to a carcinogen in the week preceding the survey: only 38 per cent of these thought their work was 'fairly bad for their health' and just 5 per cent wanted 'to change job or work because of the work hazards'. Granted, these figures are above the survey population average (28 per cent and 2 per cent, respectively), but they nevertheless clearly raise the issue of awareness or 'objectification' (Gollac, 1998) of risks.

The political mechanisms mean that involvement by employee health and safety representatives can help to give focus to demands and claims on health and safety issues, adding to the pressure on management to adopt prevention policies that work.

This political function is based on the adversarial nature of the employment relationship, in which the participants are driven by a wide range of often conflicting rationales (Coutrot, 2002). Active managerial policies (investment in ergonomics, enlisting in-house or outside expertise to identify risks and develop preventive policies) and active involvement by employees and their representatives (consultation of and action by the workers, expert participation in the prevention body, and so forth) often play into one another (Walters and Frick, 2000; Garcia et al., 2007). Where prevention is concerned, therefore, managerial activism and trade union activism more often tend to reinforce than to neutralise one another. That notwithstanding, their self-perceived economic interest often leads employers to outsource the costs of work-related health damage to the general social protection system and employees themselves, rather than pay for what may be costly prevention programmes. Employees and their representatives can therefore significantly influence the prevention policies implemented, through conflict, cooperation, or, more likely, a combination of the two.

A schematic model of the different relevant cognitive and political mechanisms has been constructed to analyse the action of HSCs

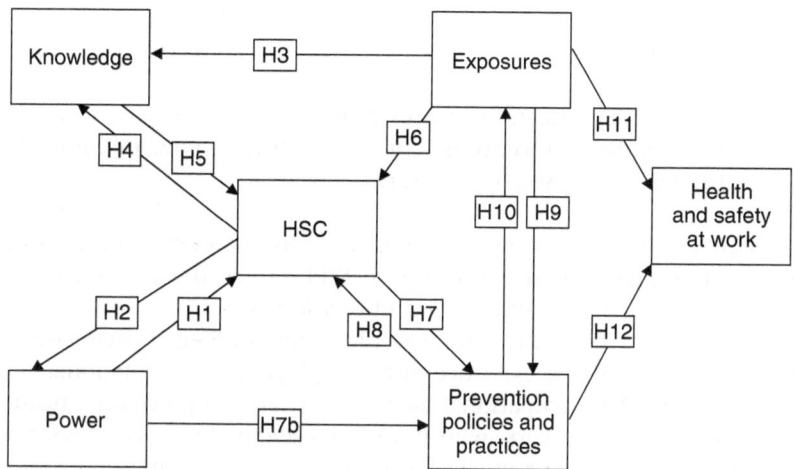

Figure 3.1 The conceptual model and its key hypotheses

(Figure 3.1) so as to account for the main interactions at work. The hypotheses underlying it are:

H1, a 'labour pressure' effect: The workforce's cohesion gives it the power to get an HSC set up and entrenched;

H2, a 'legitimisation' effect: Effective action by the HSC is apt to increase workforce cohesion and representatives' legitimacy;

H3, a 'visibility' effect: The scale of exposures fosters awareness of the hazards among the parties concerned;

H4, an 'alert' effect: The HSC's meetings and activities foster awareness of the hazards among the parties concerned;

H5, a 'concern' effect: Awareness of the hazards among management and employees promotes the creation of HSCs to develop prevention;

H6, a 'regulation' effect: The scale of exposures prompts employer compliance with the Labour Code from concern for employees' health or fear of enforcement and the financial penalties that the health and safety inspectorate or insurers (French health insurance system funds) might impose;

H7 (and H7b), a 'power position' effect: The action of the HSC (or trade unions) forces management to improve the protection of employees from the risks they are exposed to;

H8, a 'management' effect: Company management implements an official written health and safety policy that includes the creation or running of an HSC;

H9, a 'mirror' effect: Prevention policies adjust to the type of exposure;
H10, an 'elimination at source' effect: Prevention enables a reduction in exposures;
H11, an 'aetiology' effect: Exposures damage employees' health; and
H12, a 'secondary prevention' effect: prevention policies limit the health damage caused by exposures

This chapter sets out to test some of these hypotheses. In doing so it draws upon three data sets, the 2003 SUMER survey, the 2004 REPONSE survey, and the 2005 Working Conditions survey.

The SUMER survey relies on a specific feature of French occupational medicine, namely, the existence of an employer-financed national system of some 6000 occupational doctors, in charge of preventing health and safety hazards a work, who have to receive periodically (every two years) every worker for a medical examination. Small firms are surveyed by occupational doctors employed by multi-firm health and safety services (*'service interentreprise de santé au travail'*), while in large firms the health and safety department is generally firm-specific (*'service autonome de santé au travail'*). The survey has already run to two editions (1994 and 2003), and the third is being prepared in 2009. In 2003 some 1800 occupational doctors volunteered to conduct the survey on nearly 50,000 employees, on behalf of the *Ministère du travail*. The survey seeks to provide a nationally representative landscape of all occupational exposures – whether physical, organisational, psycho-social, chemical or biological – for the private sector and public hospitals. It also yields information on the employer firm, specifically whether or not it has an HSC, and on the preventive equipment that employers put at their employees' disposal for specific risks (noise, radiation, biological or chemical agents). The doctors who took part in the survey were also asked for their assessment of the quality of preventive measures for the job studied.

The REPONSE survey, a younger brother of the British Workplace Industrial Relations Survey (later Workplace Employee Relations Survey, WERS), has already run to three editions (1992, 1998 and 2004) (Amossé and Coutrot, 2008). Financed by the *Ministère du Travail* and conducted by a private institute (BVA), it is a nationally representative (for establishments above 20 employees in the private sector) workplace survey, where (in 2004) 2930 employers and 1970 employee representatives, interviewed face to face, give a fairly detailed description of personnel management devices and industrial relations at the workplace, notably about the existence and activity of the different institutions designed for workers representation – including CHSCTs. REPONSE 2004 also includes an employee postal survey, with nearly 8000 employees sampled in the 2930 establishments where managers have been interviewed, and in which workers are asked about their job perception and satisfaction. (In this chapter we fully use this feature of the survey by drawing on the three questionnaires for managers, workers representatives and workers.)

The 2005 Working Conditions Survey was conducted by Insee, the French national statistics institute, for the *Ministère du travail*, on about 18,000 workers from all sectors, including public administration and non-wage earners. It comes after a series of similar surveys made in 1978, 1984, 1991 and 1998, the two first of which inspired the European working conditions survey, conducted each five years since 1990 by the Dublin Foundation. The survey provides information on working conditions as they are perceived by workers themselves: work rhythms, physical nuisances, organisational constraints, and so on. In 2005 the survey included questions about, among other things, the resources workers have access to in order to deal with work-related risks: training, safety instructions, HSC and occupational medicine.

Using the above data sets as appropriate, the chapter considers the following hypotheses. First the H1 'labour pressure' and H8 'management effect' hypotheses; then the H6 'regulation effect' hypothesis, although the available data do not enable this hypothesis to be totally (in)validated. Then we look at whether the existence of an HSC helps improve preventive practices (H7). Finally, we examine the connection between the existence of an HSC and risk-awareness (H4).

Labour pressure and managerial policy ease the setting up and working of the HSC (H1 and H8)

If hypothesis H1 is true, a dynamic and united work force gives the employees and their representatives the power to insist that bodies for employee representation in health and safety be set up and run properly. More specifically, 'autonomous worker representation at the workplace, trade union support and effective communication between worker representatives and their constituencies' are 'pre-requisites for effective worker representation' (Walters et al., 2005, p. 118). Each of these three variables – autonomous representation, trade union support, and effective communication between representatives and their constituencies – can be quantified using the REPONSE survey. Hypothesis H8 predicts that it is easier to get an HSC set up where company management – and especially human resource management (HRM) – is professionalised and formalised.

In France, as already stated, a health and safety committee (*Comité d'hygiène, de sécurité et des conditions de travail*, or CHSCT) must be set up in workplaces that employ at least 50 workers, and it has an important role to play in workplace risk prevention. In practice, according to the 2004 REPONSE survey findings, only 77 per cent of the workplaces theoretically concerned actually have an HSC: 95 per cent of those with more than 500 employees, but only 57 per cent of those with between 50 and 100 employees.

The employee representatives questioned in the same workplaces give replies that tally with those of the employers (Figure 3.2). In small establishments (20 to 49 employees), however, almost twice as many employee representatives (30 per cent) as employers (16 per cent) report the existence of an HSC. This may be because the Labour Code provides that shop stewards – who are basically the employee representatives in small establishments – can perform the duties of HSCs.

To test H1 and H8 empirically, a Logit model was estimated to assess the influence of different factors on the probability of a workplace having an HSC (Table 3.1, model 1), and the probability of a given HSC holding more than the statutory minimum number of meetings (four a year) (Table 3.1, model 2).

Reflecting current legislation, workforce size is clearly the primary determinant of whether a workplace has an HSC. Branch of industry is a lesser factor, but on current trends the service sectors have a lower propensity to set up an HSC.

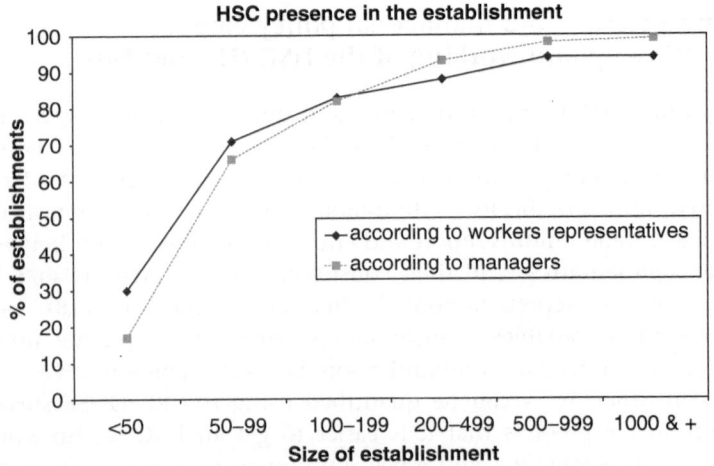

Figure 3.2 Presence of HSCs by size of workplace

Table 3.1 Labour pressure increases the probability of existence of an HSC

	Frequency	Odds ratio Model 1: Existence of an HSC	Odds ratio Model 2: More than 4 HSC meetings
	2930 workplaces	2049 out of 2930	707 out of 2049
Size of workplace			
20–49 employees	57	0.03	NS
50–99	21	0.27	0.43
100 and over (ref.)	22	–	–
Industry			
Agribusiness	3	2.27	NS
Consumer goods	4	2.61	NS
Capital equipment	6	2.30	NS
Intermediate goods	12	1.97	NS
Construction	9	NS	NS
Retail	20	1.97	NS
Transport (ref.)	7	–	–
Banking, insurance	4	NS	2.46
Business services	18	NS	NS
Personal services	8	NS	NS
Health, social welfare	9	NS	0.53
Age of establishment			
More than ten years	60	1.66	NS
Ownership and employer relations			
Membership of a group	51	1.30	1.32
Ultimate holding company of stock exchange listed group	24	1.41	NS
Company member of employers' networks	20	1.71	1.41

(continued)

Table 3.1 (Continued)

	Frequency	Odds ratio Model 1: Existence of an HSC	Odds ratio Model 2: More than 4 HSC meetings
Employee–employer relations			
Pro-trade-union management	39	NS	1.44
Presence of a union representative	53	3.42	NS
Union density <5%	59	0.56	NS
Many disputes	12	1.48	1.98
Many disciplinary measures	30	NS	0.77
Health and safety			
Repeated work accidents	11	1.44*	1.32*

Source: 2004 REPONSE survey, non-agricultural commercial sector (calculation by author)

Note: All the odds ratios presented are significant at the 5% level (except * at the 10% level only, and NS not significant)

Meaning of variables:

- 'employers' networks': The establishment regularly takes part in two outside bodies (industry federation, professional association, employers' club, board of directors of a non-group company)
- 'pro-trade-union management': The management representative considers that 'the trade unions play an irreplaceable part in representing employees'
- 'many disputes': The workplace has experienced at least two different types of dispute in the last three years (of the following eight types of dispute: petition, demonstration, walkout, go-slow, work-to-rule, refusal to work overtime, strike of fewer than two days, strike of two days and more)
- 'many disciplinary measures': Disciplinary measures applied to more than 6 per cent of employees in 2004
- 'repeated work accidents': The employer reports that the workplace has experienced 'repeated work accidents' in the last three years

The influence of managerial policies is immediately apparent, and confirms hypothesis H8. First, being part of a group increases the probability of a workplace having an HSC – even more so where the group is listed on the stock exchange. This is because such workplaces often apply the written HRM policies laid down centrally by head office, especially as regards the working of representation bodies and workplace risk prevention. Likewise, heads of establishments which have an active involvement in outside business-related organisations (industry federation, business club, board of directors of another company, and so forth) are more apt than others to set up an HSC due to having access to management information and support tools. This effect from membership of outside employers' networks is particularly found for small and medium-sized enterprises because most large firms are already part of such networks and implement written HRM policies.

The existence of an HSC therefore depends on managerial policies, but also reflects internal labour pressure, clearly validating hypothesis H1. Each of Walters' three criteria – 'autonomous representation', 'trade union support' and 'effective communication' between representatives and those who appoint them – can be matched with a REPONSE survey indicator; all exert a significant positive influence on the existence of an HSC.

'Trade union support' is ensured mainly by the presence of a trade union in the establishment or workplace. The plain fact is that HSCs are more often found in workplaces where there is trade union representation. So, among workplaces with between 50 and 99 employees, 65 per cent of those with a union representative have an HSC compared with 44 per cent of the workplaces with no union representative.

The 'autonomy' of workforces and their representatives from management can be gauged by their ability to engage in collective disputes to defend their interests in the workplace (Cézard et al., 1996). By this yardstick, the existence of an HSC correlates closely with the existence of autonomous workforces; so, among workplaces with trade union representation, an HSC is found in only 48 per cent of those that had experienced no form of collective labour dispute during the three years preceding the survey (2002 to 2004), against 64 per cent that had experienced one dispute of some kind and 74 per cent for workplaces that had experienced several. This clearly contradicts the idea – often expressed in the managerial literature – that joint consultation on health and safety operates best in the absence of industrial conflict.

Finally, the effectiveness of communication between employees and their representatives can be evaluated from employees' own responses. Sixty-three per cent agreed with the statement that 'the employee representatives do a good job representing workers' demands', and 33 per cent said they went to the employee representatives first with some complaints.[1] There is a clear correlation between this quality of communication and the existence of an HSC, which raises the preceding percentages to 65 per cent and 37 per cent. These ostensibly small differences are nevertheless highly statistically significant.[2]

Most HSCs (57 per cent) met the statutory four times in 2004. However, one in five HSCs held fewer than the statutory number of meetings, while another one in five HSCs held more. Here again, proneness to workplace dispute has a major impact on HSC activity: where there no were disputes in the three years preceding the survey, only 5 per cent of HSCs met more than four times, against 25 per cent in workplaces that had experienced several types of disputes. Factors that also increase the likelihood of HSCs meeting more than the statutory minimum (Table 3.1) include how long the workplace had been established, management's involvement in outside business bodies, and a pro-trade union attitude.

The existence of major risks favours the setting up of an HSC (H6)

The conceptual model in Figure 3.1 has no clear-cut predictive value for the connections between exposure and the existence of an HSC. It is true that hypothesis H6 and the interaction of hypotheses H3–H5 (exposure to major risks promotes hazard awareness and hence the creation of HSCs) predict that the presence of risks should favour the setting up of an HSC. But equally likely is a converse effect through the operation of hypothesis H9 (elimination of risks at source) resulting from an active policy of prevention promoted by the existence of an HSC (H7).

The only way to measure the impact the existence of an HSC may have on changes in exposure would be to use time-series data, but we only have cross-sectional data. Any observed statistical relationship between exposures and HSC existence will therefore reflect the outcome of two conflicting effects H6 and H7–H9. The lack of a statistical relationship between HSC existence and exposure does not, therefore, necessarily point to the lack of a causal link, since high exposures may have fostered the creation of an HSC whose activity may have been effective in reducing exposures.

The correlation between the existence of an HSC and occupational exposures is therefore measured cross-sectionally (Table 3.2), with the estimate controlled by a wide range of contextual variables (size, sector, composition of workforce, and so on). To summarise the very wide range of occupational exposures described in the 2003 SUMER survey questionnaire, 12 synthetic indicators of exposure to the different listed risks were constructed:

- significant exposure to at least one chemical:[3] At least one chemical with an intensity not reported as 'very low' by the company doctor (frequency = 21 per cent);
- exposure to biological agents (regardless of duration) (frequency 15 per cent);
- exposure to radiation (regardless of duration) (frequency 4 per cent);
- exposure to fast-paced work: The employee's work pace is imposed by at least three of the nine constraints listed in the survey[4] (frequency 30 per cent);
- exposure to major physical effort or strain[5] (frequency 24 per cent);
- high exposure to work on visual display units (VDUs): The employee works for more than 20 hours per week on a computer system (frequency 21 per cent);
- exposure to night work: More than 20 nights a year (frequency 12 per cent);
- exposure to shift work (2 x 8, 3 x 8, and so on) (frequency 13 per cent);
- unpredictable work schedules: The employee does not know his following week's work schedule (frequency 13 per cent);
- long working hours: 45 hours or more in the last week worked (frequency 14 per cent);
- lack of job discretion: The employee generally calls on others in the event of an incident (frequency 22 per cent); and
- insufficient co-worker support: The employee reports not being able to get assistance from others in order to do his job properly (frequency 7 per cent).

The model's outcomes (Table 3.2) point to an association between exposure to some hazards – shift work and night work, work on VDUs, radiation, an accumulation of work pace constraints – and a markedly greater frequency of an HSC. By contrast, HSCs are somewhat less frequently found where jobs entail exposure to chemical hazards and unpredictable work schedules. Overall, the combined effect of H6

Table 3.2   Some occupational exposures increase the probability of existence of an HSC

	Percentage of HSCs	Odds ratio (Logit model)
All workplaces	50	
Characteristics of the workplace		
Has an autonomous occupational health service	94	3.94
ISO standard	78	2.68
Works 35-hour week	59	4.41
Size		
<20 employees	10	0.04
20–49	24	0.06
50–499	81	ref.
500 and over	97	5.60
Sector		
Building/civil engineering	30	0.68
Hospitality, travel, audiovisual/press and personal services	29	0.68
Employee characteristics		
Male	52	1.14
Length of service <1 year	27	0.73
Fixed-term contract	34	0.81
Temporary agency worker	25	0.12
Public servant	89	0.60
Part-time	36	0.86
Job-related occupational exposures		
Long workweek (> = 45 H)	51	NS
Shift work	79	1.46
Night work (+20 nights/year)	65	1.13
Unpredictable work schedules	43	0.88
Work on VDUs (+20 hours/week)	58	1.20
High physical constraints	46	NS
Noise	54	NS
Biological agents	56	NS
Chemicals (significant exposure)	47	0.89
Radiation	66	1.27
High work-pace constraints	61	1.30
Little job discretion	48	NS
Little social support	45	NS

Source: SUMER survey 2003 (calculation by author)

'regulation' with the sequence of effects H3–H5 ('manifest' and 'concern') is what seems to prevail: managements are more apt to set up HSCs where the work-related risks are high in their workplace because of the technical and organisational characteristics of their business. But the traces of an H9 effect ('elimination at source') are to be found for chemical exposures – at least, this is what is suggested by the negative association between chemical hazards and the existence of an HSC (a relative effectiveness confirmed below). On the whole, as we might have expected, these results are not clearly conclusive in favour of any specific hypothesis among those under review.

If we set occupational exposures aside, certain workplace characteristics significantly increase the probability of an HSC: along with large workplaces, many more of those that work the reduced 35-hour week have introduced an International Organization for Standardization (ISO) standard or have an autonomous occupational health service also have an HSC. These indicators may well reflect a management propensity to fulfil labour (35 hours), organisational (ISO) or health (autonomous service) standards, which would also explain the compliance with employment regulations. But it could also be a 'management effect' in play. The SUMER survey yields no information about employer–employee relations in the workplace – internal 'labour pressure'.

HSCs are associated with a better quality of prevention (H7)

Testing hypothesis H7 means measuring the correlation between the quality of risk prevention and the existence of an HSC, again controlling the estimate by a wide range of contextual variables, including exposures. The three surveys we are drawing on in this chapter provide several original and innovative indicators for the quality of prevention, and bring new insights by distinguishing between different types of exposures.

SUMER provides two kinds of information to evaluate the quality of job-specific prevention. Whenever an exposure to risks related to noise, radiation, chemicals or biological agents is identified in the questionnaire, the doctor must specify whether the employee has personal protective equipment. The proportion of employees with this equipment varies from 45 per cent for radiation to 67 per cent for noise (Table 3.3).[6]

The odds ratios given are significant at the 1 per cent level (except *: at the 0 level, or NS, not significant). They reflect the influence of

Table 3.3   HSCs improve the quality of prevention and information

Percentage of exposed employees supplied with personal protective equipment against ...	With an HSC	Without an HSC	All establishments	Odds ratio
Noise	73	60	67	1.18*
Radiation	47	44	45	NS
Biological agents	59	35	49	1.30
Chemicals	67	57	62	1.19
Adverse doctor's assessment on ...				
(1)organisational risk	20	17	18	1.16
(2)physical risk	31	32	31	1.10
(3)chemical risk	25	35	30	0.81
(4)biological risk	12	22	17	0.83
Employee had an accident with work stoppage within the last 12 months ✿	4.1	4.6	4.4	NS
Employee considers his work to be tiring ✿	29	27	28	NS
Employee considers his work to be stressful ✿	38	32	35	NS
Employee considers his work to be harmful to his health ✿	30	24	27	1.13
Employee considers his health to be poor ✿	21	18	19	NS

Source: SUMER survey 2003 (✿ self-administered questionnaire)
Note: (1) coverage: all employees commercial sector; (2) coverage: employees exposed to a physical constraint (i.e. 77 per cent of employees); (3) coverage: employees exposed to a biological agent (i.e. 15 per cent); (4) coverage: employees exposed to a chemical (i.e. 38 per cent)

the existence of an HSC on the probability of occurrence of the event (e.g., having had a work accident), in a Logit model where the control variables are the characteristics of the workplace, employees and job (occupational exposures) (see Table 3.2). Exceptionally, the models concerning personal protective equipment do not include occupational exposures as control variables since they are estimated only on the sub-population exposed to the risk (the only one that can benefit from protective equipment).

Other information that can be used to evaluate the quality of prevention comes from a unique feature the SUMER survey – the occupational expertise of the survey-taking company doctor, who is asked to give a

Table 3.4 'Doctor's assessment of the disease risk' (per cent)

'How do you assess the quality of the job and/or the work environment in terms of ... '	Very poor	Poor	Good	Very good	No reply	All
work organisation (1)	2	17	65	14	2	100
preventing exposure to physical constraints (2)	3	29	54	8	6	100
preventing exposure to biological agents (3)	1	15	62	12	10	100
preventing exposure to chemicals (4)	3	27	51	8	11	100

Source: SUMER survey 2003
Note: (1) coverage: All employees commercial sector; (2) coverage: Employees exposed to a physical constraint (i.e. 77 per cent of employees); (3) coverage: Employees exposed to a biological agent (i.e. 15 per cent); (4) coverage: Employees exposed to a chemical (i.e. 38 per cent)

summary assessment of the 'disease risk' associated with the job, taking into account extant exposures and preventive measures. In practice, the doctor has to give a score from 1 ('very poor') to 4 ('very good') for the 'quality of the job and/or work environment' in four areas: 'work organisation', 'prevention of exposure to physical constraints', 'prevention of exposure to biological agents', 'prevention of exposure to chemicals'. The company doctors rated the quality of prevention as 'poor' or 'very poor' for a significant number of jobs (from 16 per cent for biological risks to 32 per cent for physical constraints). This is shown in Table 3.4.

Whether for personal protective equipment or the doctor's summary assessment, the results tally perfectly (Table 3.3): the presence of an HSC in the workplace is markedly associated with a better quality of prevention against chemical and biological risks. So, among employees exposed to a chemical, 67 per cent have personal protective equipment where there is an HSC, against 57 per cent where there is not; on a like-for-like basis, the presence of an HSC increases the probability of having personal protective equipment by 19 per cent, and likewise reduces by 19 per cent the probability of the doctor rating job quality as 'poor'. The figures for biological hazards are +30 per cent and –17 per cent, respectively. These results are highly statistically significant (at the level of 1 per cent). They hardly change whatever the specification of the model, whether the control variables include only workplace characteristics or also employee characteristics and occupational exposures (Table 3.5).

Table 3.5   Robustness test of explanatory models of the quality of prevention

Odds ratio of the HSC variable in a multivariate (Logit) model explanatory of 'poor job quality' in terms of ...	Model 1	Model 2	Model 3
work organisation (1)	1.17 ***	1.17 ***	1.16 ***
preventing exposure to physical constraints (2)	1.01	1.11 ***	1.10 ***
preventing exposure to biological agents (3)	0.82*	0.84	0.81*
preventing exposure to chemicals (4)	0.77 ***	0.82 ***	0.83 ***

***significant at 1% level
*significant at 10% level
Source: SUMER survey 2003
Note: (1) coverage: all employees; (2) coverage: employees exposed to a physical constraint (i.e. 77 per cent of employees); (3) coverage: employees exposed to a biological agent (i.e. 15 per cent); (4) coverage: employees exposed to a chemical (i.e. 38 per cent)
Model 1: Control variables = workplace characteristics only
Model 2: As 1 + employee characteristics
Model 3: As 2 + exposure to work-related risks

The findings for noise tell the same story: where there is an HSC, 73 per cent of the employees have protective equipment against 60 per cent where there is not. But the connection is less certain where other factors are equal: the odds ratio (1.18) is significant only at the level of 10 per cent.

The positive association between HSC's presence and the quality of prevention policies is confirmed by the Working Conditions Survey (Coutrot, 2008). On a like-for-like basis (workplace size and sector, employee and job characteristics), the presence of an HSC more than doubles the probability of an employee having been given training or information on occupational risks within the last 12 months. It also more than doubles the probability of the employee having been given written safety instructions (Table 3.6). Finally, the existence of an HSC is associated with a 20 per cent increase in the probability that the employee will declare to follow on all points the written instructions given to him.

The odds ratios given are significant at the 1 per cent level (except NS, not significant). They reflect the influence of the existence of an HSC on the probability of occurrence of the event (for example, having been given information on risks) in a Logit model where the control variables are the characteristics of the workplace, employees and job (occupational exposures).

Table 3.6 HSCs improve the quality of prevention and information

Percentage of employees who have ...	Workplace with an HSC	Workplace without an HSC	All establishments	Odds ratio
been given *information* on risks within the last 12 months	25	12	19	2.01
been given *training* on risks within the last 12 months	29	9	20	2.72
been given written safety instructions	57	25	43	2.68
followed the instructions in full (for those given them)	65	60	63	1.21
had an accident with work stoppage during the last 12 months	8.5	8.2	8.4	NS
do not think they will able to do the same job at the age of 60	39	42	40	0.91

Source: 2005 Working Conditions survey, INSEE-Dares
Note: Coverage: All employees, all sectors

HSCs improve risk awareness (H4)

By contrast, the presence of an HSC is significantly associated with an increase in the probability of the doctor rating the job quality as poor in terms of work organisation (+16 per cent) and physical risks (+10 per cent). Likewise, it adds 13 per cent to the probability that employees will judge their work to be harmful to their health (see Table 3.7). Where other factors are equal, however, it is neutral on the probability of having suffered a work accident with work stoppage within the past 12 months, as well as self-reporting poor health,[7] or that the work is 'tiring' or 'stressful'.

What explanation can be advanced for this greater concern by doctors and employees where there is a workplace HSC? It was seen that HSCs are more frequently found in workplaces where exposure to physical and organisational risks is high (see Table 3.2). But exposures are one of the control variables (Table 3.7), and a major determinant of the doctor's concern: for example, unpredictable work schedules increase

Table 3.7 Econometric analysis of the doctor's assessment of the quality of prevention against the different risks (Logit model)

Poor prevention of ...	Physical risks	Organisational risks	Chemical risks	Biological risks
Workplace characteristics				
Has an 'autonomous' occupational health and safety service (vs multi-firm Health and Safety Service)	0.83	0.74	0.76	NS
Follows ISO standard	0.78	0.80	0.85	0.82
Applies 35-hour week	0.88	0.89	0.90	NS
Has an HSC	1.10	1.16	0.83	0.81*
<20 employees	0.80	0.76	NS	1.33
20–49	NS	NS	NS	1.32
50–499	ref.	ref.	ref.	ref.
500 and over	NS	NS	0.72	0.78
Sector	NS	–	NS	NS
Construction	NS	0.70	NS	NS
Employee characteristics				
Male	0.72	0.84	1.17	1.19
Length of service <1 year	NS	NS	NS	NS
Fixed-term contract	NS	NS	NS	NS
Temporary agency worker	NS	NS	NS	NS
Non-EU foreigner	1.43	1.31	NS	NS
Part-time	NS	NS	0.87	NS
Occupational exposures				
Long workweek (> = 45 H)	NS	1.81	NS	NS
Shift work	1.17	1.10	NS	NS
Night work (+20 nights/year)	NS	1.25	0.79	NS
Unpredictable work schedules	1.15	1.62	NS	NS
Work on VDUs (+20 hours/week)	0.77	1.12	NS	NS
High physical constraints	1.30	1.16	1.17	NS
High noise	1.44	1.21	1.45	1.36
Biological agents	1.27	1.17	0.87	*
Chemicals (significant exposure)	1.38	1.11	2.10	1.28
Radiation	1.18	1.19	1.14	NS

Table 3.7 (Continued)

Poor prevention of ...	Physical risks	Organisational risks	Chemical risks	Biological risks
High work pace constraints	1.33	1.67	1.13	1.20
Little job discretion	NS	1.11*	NS	NS
Little co-worker support	1.66	2.64	1.29	1.28
N	38,281 (77%)	49,855 (100%)	18,904 (38%)	7,692 (15%)
coverage of the model	Workers with at least one physical risk	Complete sample	At least one chemical exposure	At least one biological exposure

*All the odds ratios are significant at the level of 1% except (at the level of 10%)
Source: SUMER survey 2003
Interpretation: For identical workplace, employee and job characteristics, there is a 17 per cent lower probability (odds ratio = 0.83) of an occupational doctor who belongs to an 'autonomous' occupational health service reporting a risk of health damage from physical risks than of a doctor who belongs to a multi-firm service. Conversely, when there is an HSC in the workplace, the probability of reporting a risk of health damage from physical constraints is 10 per cent higher. The industry has no significant influence, except construction for organisational risks.

by 62 per cent the probability that the doctor will rate the job as of poor quality in organisational terms. Exposure is therefore not the only thing that causes doctors and employees more concern. Counter-intuitively, might the presence of an HSC actually lower the standard of prevention against physical and organisational risks? Theoretically, that seems unlikely, and indeed no observable connection is found between the existence of an HSC and work accidents or perceived poor health. Employees covered by an HSC are more concerned about the health impact of their work, but they are no less healthy than other employees (on a like-for-like basis). It is not, therefore, that HSCs increase the actual risks; rather, they increase the participants' aware-ness of them (H4 effect). The HSC's activities, the discussions at its meetings, the investigations it can carry out give a prominence to risks that might otherwise go unperceived.

Unlike for chemical or biological risks, there is no other preven-tive measure for physical and organisational risks than elimination (or attenuation) at source: no protective measure to attenuate the risk can be interposed between the cause of harm and the employee.

The SUMER questionnaire therefore includes no questions about protection against physical or organisational risks that would (in the same way as for chemical and biological risks) enable a more objective assessment of HSCs' impact on prevention. It can only be said that HSCs are more commonly found in workplaces where organisational risks are high (see Table 3.2): the results described here seem to suggest that HSCs do not significantly reduce physical, organisational and psychosocial risks but arguably are effective where chemical and biological risks are present.

Conclusion

The study's findings obtained from the use of different types of indicator and statistical sources are consonant with the data in the international scientific literature: the presence of employees' health and safety representatives (HSCs) is clearly associated with better prevention policies at the workplace level. This is particularly noticeable in view of the statistical robustness of the results, given the size of the samples we could mobilise and the diversity of the indicators used to describe the quality of prevention, including several that have scarcely or never been used in the international literature. The provision of individual protections against exposures, the provision of health and safety training or written safety instructions, the judgment of occupational doctors about the quality of prevention: all of these indicators are unambiguously better where HSCs are present. We also have been able to show that this association between HSCs' presence and the quality of prevention depends strongly on the nature of occupational exposures: it is quite clear for chemical or biological risks, but is not to be found for physical or organisational constraints. We have further shown a positive association between HSCs' presence and workers' awareness of occupational risks.

By contrast, no direct link could be made here between the presence of an HSC and employee health. Specifically, the presence of an HSC has no perceptible effect on the rate of work accidents reported by employees. This finding is not rare in the scientific literature (Nichols, Walters and Tasiran, 2007, although their specific study on WERS shows some influence of HSCs with union members on the rate of accidents). It may be due in particular to the fact (documented here) that HSCs are more frequently found where work-related risks are numerous, therefore making it difficult to report any prospective beneficial effect of HSCs on accident rates from cross-sectional data. The lack of a link between

HSCs and work accidents is borne out here even where occupational exposures are taken into account. But again, that could be explained by a favourable impact of HSCs on such exposures (H10 effect), a hypothesis that could not be tested for want of time-series data. More generally, the static nature of the available data prevented all the inter- actions described by the model presented here (Figure 3.1) from being tested to complete satisfaction, and this is a major limitation of this study. Likewise, the lack of any particulars (in the SUMER and Working Conditions surveys) of trade union presence and membership rates meant that the general impact of employee–employer relations on the quality of prevention policies could not be studied.

As well as promoting prevention policies, HSCs clearly help build participants' awareness and improve risk identification. HSCs also substantially improve prevention of chemical and biological risks. On the other hand, they appear less effective with regard to physical and organisational risks. Shop stewards presumably have fewer difficulties in demanding the implementation of protection against specifically located risks than in calling into question management and work organisation approaches.

Notes

1. More specifically, 33 per cent of the workers said they 'went first to the employee representatives to try and find a solution' for at least one of the four kinds of issue cited in the questionnaire (work conditions, lack of promotion, disagreement with a superior, dismissal procedure).
2. This is demonstrated by a Logit model (not reproduced) calculated only on workplaces for which employee responses were given, and explaining the existence of an HSC by the same variables as at Table 3.1, to which two indicators for the quality of communication between employees and repre- sentatives resulting from the 'employee' questionnaire were added. The odds ratios of the variables concerned are 1.35 (for the 'employee turns first to the employee representatives' variable) and 1.19 (for the 'worker considers that the employee representatives do a good job representing workers' demands' variable).
3. The company doctor who completes the 'Chemicals' section of the SUMER questionnaire is required to specify the intensity of the exposure for each of the 80 listed chemicals ('estimated very low', 'measured very low', 'estimated low', 'measured low', 'estimated high', 'measured high', 'estimated very high', 'measured very high'). This paper considers those exposures of an intensity not estimated or measured as 'very low', that is, just over half of all exposures.
4. The nine pace constraints are: automatic movement of a part or automatic pace of a machine; other technical constraints; immediate dependence on fellow workers' work pace; production target or deadline to be met in one

hour or less; target or deadline to be met in one day or less; external demand requiring an immediate response; external demand not requiring an immediate response; permanent control or supervision by a superior; computerised control or monitoring.

5. The employee is exposed to at least two of the following six strains: standing position for more than 20 hours per week; walking for more than 20 hours per week; repetition of the same movement for more than 20 hours per week; working kneeling for more than two hours per week; working with arms above the head for more than two hours per week; work in a twisted or squatting position for more than two hours per week.

6. This concerns the share of employees exposed to the risk (for example, chemical) with at least one item of personal protective equipment. This indicator arguably overestimates the proportion of employees who are effectively protected: there is no certainty that the personal protective equipment supplied to the worker is appropriate (workers may be supplied with skin protection where the risk is mainly by inhalation), nor that it is actually and properly used. Also, an employee who is exposed to several chemicals but is protected against only one of them will be classed as having protection in the survey.

7. For these purposes, 'poor health' is health that is self-scored as 6 or under (on a scale of 1 (very poor) to 10 (very good), that is, 19 per cent of employees).

4
Characteristics, Activities and Perceptions of Spanish Safety Representatives

Ana M. García, Maria J. López-Jacob, Isabel Dudzinski, Rafael Gadea and Fernando Rodrigo

In Spain, under the Law for the Prevention of Occupational Health Risks (1995), all workplaces with six or more workers should have one or more safety representatives. Previous Spanish legislation did not contemplate any structured arrangement for workers' participation in decisions related to occupational health and safety at workplace level.

This chapter presents the main results of a study evaluating Spanish safety representatives' activities, their perceptions of occupational risk management as well as their needs regarding training and the obstacles and supports they perceive to performing their duties. It is the first systematic investigation of such issues in Spain. It shows that, while safety representatives are active in a number of ways in relation to health and safety, the extent to which they are consulted on health and safety arrangements remains limited in many workplaces, and a significant number of safety representatives report that some compulsory activities for occupational health and safety protection in their companies (such as risk evaluation, planning of preventive interventions, and workers' consultation and participation in evaluation and planning) are not properly implemented.

Safety representatives in Spanish workplaces

In Spain occupational health and safety regulations have experienced substantial change during recent years. Although there have been laws and regulations related to health and safety at work in Spain since 1900, Law 31/1995 for the Prevention of Occupational Health Risks, derived from the EU Framework Directive 89/391, requires employers, for the first time, to manage health and safety in a systematic, informed and

73

participative way. For the first time, too, workplaces with six or more workers should have safety representatives (to be designated and elected by the employees, usually from among workers' representatives). However, the coverage of this model is not complete. According to the *Fifth National Survey on Working Conditions*, carried out in 2003, 45 per cent of companies with six or more workers had not designated a safety representative. This problem affected small and medium enterprises particularly. In some Spanish provinces, regional safety representatives are also contemplated as an alternative model, mostly for small enterprises. The duties and rights of safety representatives under Spanish law are summarised in Table 4.1.

From the time that safety representatives were first established in Spain, no systematic information has been available regarding their profile, activities, or perceived constraints on and supports for the implementation of their duties. The first such survey was undertaken

Table 4.1   Duties and rights of safety representatives in Spanish workplaces (Law 31/1995 for the Prevention of Occupational Health Risks)

Duties

- Control of occupational health and safety legislation implementation
- Promotion of workers' cooperation
- Collaboration with employers on decision making about
  - Working equipments, environment and conditions
  - Organisation of preventive activities
  - Preventive actions
  - Procedures for workers' information and training
  - Procedures of documentation regarding:
    Preventive plans
    Occupational risks evaluation
    Planning of preventive activities
    Health surveillance
    Occupational injuries and diseases
- Elaboration of reports answering employers' consultations
- Respect of confidentiality
- Participation and implementation of Health and Safety Committee activities

Rights

- To be consulted by employers regarding health and safety decision making
- To accompany health and safety technicians during risk evaluations
- To accompany labour inspectors visiting the workplace
- To be informed by Labour Inspectors about the results of their visits
- To ask for inspections from the Labour authorities

Table 4.1 (Continued)

- To have free access to occupational health and safety documents and reports affecting the enterprise
- To be informed by employers about occupational injuries and diseases
- To be informed by employers about any other health and safety-related information
- To visit working places and to control working conditions in the enterprise
- To make preventive recommendations to the employer and Health and Safety Committee
- To receive justification by the employer regarding recommendations not implemented
- To ask for unsafe work activities to be stopped
- To receive from the employer training and resources for the implementation of their duties
- To have special consideration of working time for the implementation of their duties
- To have employment and professional promotion guaranteed

by the Trade Union Institute for Work, Environment and Health (ISTAS, *Instituto Sindical de Trabajo, Ambiente y Salud*),[1] and is the source of the findings presented in the following chapter.

Objectives of the research

The study 'Analysis of tasks and perceptions of safety representatives in Spain' was carried out in 2004 with the aim of gaining knowledge on the following subjects:

- personal and occupational characteristics of safety representatives in Spain;
- activities of safety representatives in the implementation of their duties as occupational health and safety representatives;
- occupational risks perceived by safety representatives in their workplaces;
- perceptions of safety representatives regarding health and safety management in their workplaces;
- obstacles and supports perceived by safety representatives in the implementation of their duties;
- perceived situation and expectations of safety representatives regarding their information and training for the implementation of their duties; and
- perceptions of safety representatives regarding available trade union resources for the implementation of their duties.

Methods

In order to accomplish the objectives above, two complementary approaches were developed: a qualitative study with discussion groups and a survey of a national representative sample of safety representatives.

The qualitative study

Twelve discussion groups were planned with safety representatives from different sectors, from different-sized workplaces and including a women-only group. Eleven of these were successfully completed with the characteristics detailed in Table 4.2.

Discussion groups were organised and recruited by local staff from the Spanish CC.OO. trade union.[2] All the participants in the groups were representatives from CC.OO. They were all guaranteed confidentiality and anonymity for the information that they provided, and they participated on a fully voluntary basis. The group discussions were conducted in CC.OO. local premises. Discussion groups lasted between 1 hour 15 minutes and 1 hour 45 minutes. All the group discussions took place in October and November 2004.

Discussion was stimulated and guided through four main questions which were put to participants:

- Why did you become safety representatives?
- What do you like the most from your work as safety representatives?

Table 4.2   Characteristics of discussion groups and participants

Code	Location	Sector	Workplace size (workers)	Participants (m = men; w = women)	Age (years)	Time as HSR (years)
DG1	Barcelona	Industry	30–100	8 (m)	30–60	1–10
DG2	Almería	Agriculture	N.A.	3 (m) 1 (w)	23–55	1–15
DG3	Madrid	N.P.	N.P.	7 (w)	30–55	N.A.
DG4	Madrid	Services	<30	2 (m) 4 (w)	25–50	0–3
DG5	Valencia	Services	>100	9 (m) 3 (w)	30–55	1–9
DG6	Valencia	Construction	<30	7 (m)	30–55	N.A.
DG7	Sevilla	Industry	>100	6 (m)	30–57	>1
DG8	Sevilla	Construction	N.A.	7 (m)	35–55	1–10
DG9	Sevilla	Administration	N.A.	6 (m) 1 (w)	40's	1–10
DG10	Sevilla	Services	>50	1 (m) 3 (w)	30–55	N.A.
DG11	Sevilla	Services	>100	1 (m) 5 (w)	25–50	0–10

Note: HSR: health and safety representative; N.A.: not available; N.P.: not pertinent: as this group was intended to include only women, working sector was not considered.

- What is it that you least like about your work as safety representatives?
- How do you think your duties as safety representatives can be improved or made more effective?

Experienced social researchers were in charge of conducting, recording and transcribing the discussions and preparing a final report of the qualitative study in which selected transcriptions of opinions expressed in the groups were analysed.

The survey

A special questionnaire was designed. Partially based on a questionnaire previously used with trade union health and safety representatives (Garcia et al., 2005), it was revised by ISTAS and CC.OO. technicians and tested in a small pilot study (n = 9). Several different versions of the questionnaire were developed before the final one used in this research.

The questionnaire was structured in five main sections: personal data, activities, conditions and attitudes in the workplace, information and training, and resources and supports. Most of the items were answered through closed questions, including Yes/No, and three degree scales for agreement (A lot/Something/A little-Nothing), intensity (High/Moderate/Low-Null) and frequency (Always-Almost always/Sometimes/Almost never-Never). Spanish and English versions of the questionnaire are available on request from the authors.

In Spain, there is no centralised register with information on all the elected safety representatives in the country. Our only chance to have a national level representative sample was through trade union registers. These registers contain the results of elections in workplaces in which there are unionised workers. Trade union databases were therefore used in order to select safety representatives to be interviewed in this study. In these databases information from trade union elections in workplaces is registered, including location data for elected representatives. It was decided to obtain a total sample size of 1200. According to available data, this sample accounts for almost 1 per cent of all safety representatives in Spain. The sample was designed to include 60 per cent of CC.OO. safety representatives and 40 per cent of safety representatives from other trade unions and the non-unionised, this also being in accord with available information on the distribution and characteristics of workplace representatives in Spain. The sampling process was also designed to include a sufficient number of safety representatives from the different main activity sectors (agriculture, industry, construction, services and public administration) and from

Table 4.3   Expected and (actual) survey responses

	Administration	Agriculture	Construction	Industry	Services
CC.OO.					
≤30 workers	36 (36)	36 (36)	36 (36)	36 (36)	36 (36)
31–50 workers	36 (46)	36 (7)	36 (39)	36 (36)	36 (36)
51–100 workers	36 (42)	36 (5)	36 (24)	36 (36)	36 (36)
>100 workers	36 (40)	36 (14)	36 (18)	36 (36)	36 (37)
Other					
≤30 workers	24 (24)	24 (24)	24 (24)	24 (24)	24 (24)
31–50 workers	24 (14)	24 (4)	24 (21)	24 (24)	24 (24)
51–100 workers	24 (14)	24 (5)	24 (15)	24 (24)	24 (24)
>100 workers	24 (20)	24 (5)	24 (8)	24 (24)	24 (24)

workplaces with different sizes (<30, 31–50, 51–100 and >100 workers). The distribution of expected number of interviews according to these requirements is shown in Table 4.3 (with actual numbers in brackets).

Personal phone interviews were carried out by a private company. A list with the names and contact address of selected safety representatives was provided, together with up to four substitutes for each index subject included with the same characteristics regarding trade union, sector and workplace size.

The mean time required for each interview was 24 minutes. Interviews were carried out with the Computer Assisted Telephone Interview (CATI) technique and were conducted from September to December 2004.

Results

General characteristics of Spanish safety representatives

The final sample included 68 per cent of CC.OO. representatives (n = 824), 21 per cent of representatives from other unions (n = 249) and 11 per cent non-unionised safety representatives (n = 127), suggesting that CC.OO. and other unionised representatives could be slightly over-represented. Details of the characteristics of the safety representatives surveyed are presented in Table 4.4.

The discussion groups suggested that many people became safety representatives by chance, although some of them also report previous interest in safety and health issues:

I liked the subject and there were no other volunteers (DG5).

Women tended to face additional obstacles:

I have pressure, but a lot of it, for being a woman (DG3).

Table 4.4 Characteristics of Spanish safety representatives. National survey, 2004 (n = 1201)

	(n) %
Sex	
Men	(912) 75.9
Women	(289) 24.1
Age (years)	
16–25	(25) 2.1
26–35	(304) 25.3
36–45	(445) 37.0
46–55	(354) 29.5
56–70	(73) 6.1
Contract	
Temporary	(73) 6.1
Permanent	(1128) 93.9
Time in the company	
≤5 years	(229) 19.1
6–10 years	(287) 23.9
11–15 years	(213) 17.7
16–20 years	(170) 14.1
>20 years	(300) 25.0
Unknown	(2) 0.2
Time as safety representative	
0–1 years	(347) 28.9
2–3 years	(387) 32.2
4–6 years	(270) 22.5
>6 years	(192) 16.0
Unknown	(5) 0.4

In the survey, safety representatives mostly report high levels of interest in occupational and safety issues (72 per cent), moderate levels of training (68 per cent), moderate levels of experience (57 per cent) – although mean time as safety representative is relatively high (four years) – and moderate levels of satisfaction with the implementation of their duties (55 per cent). Reported level of interest increases with workplace size, but this characteristic seems not to be related with level of satisfaction with their duties (Table 4.5).

Safety representatives from agriculture and construction reported the highest satisfaction with the implementation of their duties, while representatives from public administration reported the lowest.

The discussion groups indicated that a key cause of dissatisfaction was the criticism directed at them from both the employers' and workers' sides:

> to be forced to act every day with the company against you and with the workers against you (DG1)

Table 4.5 Level of interest and satisfaction with the implementation of their duties in Spanish safety representatives. National survey, 2004 (n = 1201)

Company size (workers)	'High' interest in health and safety issues (n) %	'High' satisfaction regarding health and safety duties (n) %
≤30	(292) 65.8	(133) 29.9
31–50	(176) 71.5	(75) 30.5
51–100	(167) 71.4	(66) 28.2
>100	(231) 83.7	(88) 31.9
	(p < 0.001)	(p = 0.948)

We are the villains, always, for one of them and for the other of them, we are caught in the middle (DG4).

Satisfaction with being a representative increases with time, related to the capacity to solve workmates' problems and to force employers to do things better. Of those with less than one year's experience as a safety representative, 23 per cent were highly satisfied; of those with one to three years, 30 per cent were; of those with more than three years experience, 35 per cent were.

Activities carried out by Spanish safety representatives

Table 4.6 records the activities developed by safety representatives during the year before the interview. For the total sample, the mean number of activities relating to the provision of information and advice (minimum 0, maximum 10) was 6.0 (standard deviation, SD 2.1), with 1 per cent of the interviewees having not developed any activity in this group. The mean number of activities relating to participation in occupational health management (minimum 0, maximum 9) was 4.1 (SD 2.6), with 11 per cent of the interviewees having not developed any activity of this group. The mean number of activities relating to pressure and negotiation (minimum 0, maximum 6) was 2.4 (SD 1.3), with 11 per cent of the interviewees having not developed any activity of this group. After transformation the three groups of activities into a 0–10 ranged scale, the means (95 per cent CI) for information, management and pressure actions were, respectively, 6.8 (6.7–6.9), 4.5 (4.4–4.7) and 4.0 (3.8–4.1).

More frequent activities did not vary substantially by sector, company size or time as safety representative. Frequency of activities developed

Table 4.6 Activities performed by Spanish safety representatives during the year before the interview. National survey, 2004 (n = 1201)

	n	(%)
Information and advising		
Answering workers' consultations	1084	(90.3)
Visiting workplaces	954	(79.4)
Examining available documentation on OHS in the company	895	(74.5)
Workers' information and/or training	888	(73.9)
Meetings or consultations with own trade union	766	(71.4)[a]
Asking workers for information on OHS problems	839	(69.9)
Meetings and conversations with workers	769	(64.0)
Consultations with occupational health and safety service	747	(62.2)
Joint activities on OHS issues with other unions	285	(23.7)
Participation in OHS management in the company		
Participating in risk assessment	787	(65.5)
Participating in prevention planning	730	(60.8)
Accompanying prevention technicians for risk assessments	707	(58.9)
Answering requirements from employers on OHS issues	664	(55.3)
Participating in preventive activities related to temporary workers	473	(39.4)
Participating in accident investigation	441	(36.7)
Accompanying labour inspectors during their visits	409	(34.1)
Answering requirements from employers on environmental issues	336	(28.0)
Participating in preventive activities related to external workers	334	(27.8)
Negotiation and pressure actions		
Reporting OHS problems to supervisors/managers	917	(76.4)
Attending Health and Safety Committee meetings	729	(60.7)[b]
Participating in collective agreements	679	(56.6)
Submitting a proposal to stop unsafe work activities	220	(18.3)
Reporting to the labour inspection OHS violations in the company	171	(14.2)
Protest or complain actions (demonstrations, etc.) with workers	133	(11.1)

OHS: occupational health and safety

[a] Calculated only for unionised safety representatives (n = 1,073)

[b] In Spain only workplaces with 50 or more workers are obliged by law to form Health and Safety Committees

was higher in larger companies and also higher for safety representatives with more years of experience.

In discussion groups some safety representatives expressed the view that they performed tasks that were not their responsibility, but that of the employers, such as workers' training and providing information about occupational risks. Also, they wanted more information to be given to workers regarding their duties and activities as safety representatives, which were frequently unknown or misunderstood. Although safety representatives are legally required to be consulted by employers regarding health and safety issues, this is relatively uncommon, a point that was also reported repeatedly in discussion groups:

> Nobody there has asked me about what risks do I see or how could those risks be controlled (DG4).

In Table 4.7 the mean numbers of activities for each of these groups of activities are compared according to the different characteristics of

Table 4.7 Mean number of activities carried out by Spanish safety representatives, National survey, 2004 (n = 1201)

	Information	Management	Pressure
Gender			
Men	6.8	4.7	4.0
Women	6.7	4.0	3.7
	$p = 0.345$	$p = 0.001$	$p = 0.009$
Age			
≤ 40 years	6.7	4.4	3.9
> 40 years	6.9	4.7	4.0
	$p = 0.165$	$p = 0.055$	$p = 0.298$
Time as safety representative			
≤ 3 years	6.6	4.1	3.9
> 3 years	7.1	5.2	4.1
	$p < 0.001$	$p < 0.001$	$p = 0.091$
Sector			
Agriculture	6.5	4.7	3.4
Construction	6.3	5.0	3.7
Industry	7.2	5.2	4.5
Services	6.6	4.2	3.9
Administration	6.9	3.7	3.8
	$p < 0.001$	$p < 0.001$	$p < 0.001$
Workplace size			
≤ 30 workers	6.0	3.8	3.0
31–50 workers	6.7	4.7	3.9
51–100 workers	7.0	4.8	4.3
>100 workers	7.9	5.2	5.3
	$p < 0.001$	$p < 0.001$	$p < 0.001$

Activities grouped for information, occupational health management, negotiation and pressure actions (see Table 4.6) on a scale of 0–10.

interviewees. As seen in this table, some groups of activities are more frequent for male, older and more experienced representatives. However, the greatest differences are seen for economic sector and workplace size.

Obstacles and supports

Employers' attitudes towards safety representatives are generally valued positively: more than 75 per cent of interviewees think that employers help them in the implementation of their duties, allow access to relevant documentation and are prone to negotiate with them. But only 57 per cent of interviewees rate employers positively for implementing safety representatives' recommendations. By contrast, the attitudes of other agents involved in prevention were generally well evaluated, the proportion of safety representatives declaring they do not feel properly supported by them ranging from 16 per cent for supervisors and 12 per cent for occupational health services to 9 per cent for workers and 5 per cent for labour inspectors.

However, in discussion groups greater disagreement was stated regarding cooperation from these different agents:

> You say to a workmate that the company has put money into a protection measure and he says: 'I would prefer an increase in my salary' (DG5).

> They are worse, the workmates than the bosses. It's true. At least in my company (DG3).

But some safety representatives see that workers' attitudes are also strongly determined by company attitudes:

> If workers always found it [protection measures] there would arrive a time when they asked themselves for it when they didn't find it (DG6).

> You ask a worker: 'why are you working like this?' And he says: 'because I'm paid by the metre, and I can't work in a different way' (DG2).

Employers are also blamed for promoting bad relationships between workers and safety representatives:

> They [employers] try to put you against your workmates (DG4).

Negative attitudes from supervisors and prevention technicians were also reported in discussion groups. Besides, safety representatives demanded more support from labour inspectors.

Table 4.8   Perceived support from trade unions by Spanish safety representatives. National survey, 2004 (n = 1073 unionised safety representatives)

	High degree of agreement (n) %
My trade union adequately supports my training	(640) 60
My trade union gives me the information I need	(685) 64
My trade union complies satisfactorily with my requests	(753) 70
My trade union comes to my company if I need it	(900) 84

Generally, safety representatives evaluate their own trade union positively, with slightly less agreement regarding training and information (Table 4.8). However, perception of support from the trade union varies according to company characteristics, being significantly lower for agriculture, construction and services and also decreasing with company size.

Lack of time is one of the major obstacles pointed out by interviewees. Only 37 per cent of safety representatives think they have enough time to develop their duties adequately. In discussion groups the shortage of time and workmates misunderstanding the time that safety representatives had available was also commented on.

A large majority of safety representatives have received training in occupational health and safety (78 per cent). Safety representatives from construction and industry have received training more frequently than representatives from other sectors. However, while in industry most of the training was received from trade unions, in construction it was mostly received from other sources, especially from the company itself. Safety representatives in agriculture exhibited the lowest frequency of training (65 per cent, mostly from the trade union). Training is also much more frequent in large companies (with more than 100 workers, 93 per cent of safety representatives have received training) than in small ones (with 30 or fewer workers, only 70 per cent). Training was related to satisfaction in the implementation of their duties: 68 per cent of those feeling 'low' satisfaction vs 83 per cent of those feeling 'highly' satisfied had received training.

Most safety representatives rated their knowledge of occupational health and safety as 'moderate' (68 per cent) or 'low' (13 per cent). It should also be noted that the desires for more 'training' and 'information' were the most frequent demands from safety representatives to their trade union. The need for this was also expressed in discussion groups. Safety representatives felt overwhelmed with technical difficulties related to risk prevention:

> My problem is that there are too many technical terms in the prevention plans presented by the company. ... We don't have the knowledge of the occupational technicians developing these plans (DG5).

And more training was also demanded (although in the survey most claimed to have received more than 30 hours of training):

> It should be some kind of continuous training (DG6).

Risk perception

The risks most frequently reported by safety representatives are presented in Table 4.9. Ergonomic and organisational risks were the most common in all sectors and sizes of workplace, with even the risk of accidents in sectors such as construction considered of lower priority than other health problems.

Health and safety management in the company

Most of the safety representatives considered that the health and safety of workers in their companies was sufficiently protected (71 per cent). This positive perception was higher in the construction and agriculture sectors and in smaller companies. It was also high for males, for safety representatives with more years of experience, among those in trade unions other than CC.OO., and among those with lower interest in health and safety issues and with higher satisfaction as safety representatives. Level of training in occupational health and safety was not related to positive or negative perception regarding health and safety protection in the workplace. Despite this, in discussion groups several critical points were expressed regarding health and safety bureaucracy, small companies and economic issues:

> In my company we lack everything ... everything is right on paper (but) they fulfil nothing (DG3).
>
> In small companies there are no protection systems (DG1).
>
> They say we are going to ruin the company. But the company is ruining our lives (DG1).

When answers evaluating general protection of health and safety in the companies are distributed according to answers to specific problems related to health and safety management, some interesting

Table 4.9   Perceived occupational risks by Spanish safety representatives in their workplaces. National survey, 2004 (n = 1201)

By sector	More frequently reported risks	By size (workers)	More frequently reported risks
Construction	Awkward postures Heavy lifting Pace of work	≤30	Awkward postures Monotonous work Pace of work
Agriculture	Awkward postures Movements Temperature	31–50	Awkward postures Movements Pace of work
Industry	Movements Awkward postures Noise	51–100	Awkward postures Movements Pace of work
Public administration	Awkward postures Pace of work Physical load	>100	Awkward postures Movements Pace of work
Services	Awkward postures Pace of work Movements		

relationships appear. Negative perception of workers' health and safety protection is higher among safety representatives from companies not investing in health and safety, from companies having not carried out compulsory risk assessments or having not included all relevant occupational risks in their assessment, and from companies where workers or their representatives have not participated in risk assessment or in planning preventive actions

Discussion

The rationale of participation of workers and their representatives in the management of occupational health and safety (OHS) is twofold (Frick and Sjöström, accessed 2006). On the one hand, workers and representatives are able to help employers and their managers to identify and prioritise OHS problems and to develop and implement proper preventive measures. On the other hand, there is a conflict of interests between workers (primarily demanding healthy workplaces) and employers (primarily looking for cost-efficient production). This conflict demands interaction and consensus among workers and employers, and should be supported by legislation and surveyed and reinforced by public authorities.

The more frequent activities developed by Spanish safety representatives, related to information and advising, are necessary but not sufficient actions for real participation of workers in decisions concerning their health and safety. Participation in OHS management activities is more likely to reflect real involvement of safety representatives in relevant OHS decisions. In our sample, frequency of activities related to participation was lower than frequency of activities related to information and advising. The main supports for participation in OHS management activities were perceived to come from employers, occupational health and safety services and labour inspection. Positive management attitudes towards and implication in OHS has been repeatedly argued to be a necessary condition for effective workers' participation (Walters and Gourlay, 1990; Milgate et al., 2002; Walters et al., 2005). Empirical data have shown that perceived employer commitment to OHS in the company improves workers' participation and behaviours towards health and safety (Simard et al., 1999; Garcia et al., 2004).

In our data we found a positive association between representatives' participation in OHS management and perceived support from OHS services. The need to require employers to use competent preventive services and to involve worker representatives in decisions concerning their use has been frequently pointed out (Walters, 2006).

Pressure actions were developed by our interviewees with the lowest frequency. It has been stated that workers largely use arguments and resources for OHS more in a cooperative than in a conflict-oriented interaction with management (Frick and Sjöström, accessed 2006); for example, the right to stop dangerous work is used only as an ultimate instrument, although it has been claimed as a necessary symbolic power to strengthen safety representatives' influence (Walters, 2006). Eighteen per cent of our interviewees reported having submitted a proposal to stop unsafe work, using their rights established in Spanish legislation. Perceived labour inspection support was shown to be a positive reinforcement for every group of activities. Safety representatives are asking for greater support from public authorities. Unfortunately, labour inspectors are generally not engaged with ensuring the proper operation of worker representation in workplaces in Spain or elsewhere (Walters et al., 2005; Frick and Sjöström, accessed 2006).

Interestingly, our results show that trained safety representatives are more active than untrained ones (Garcia et al., 2007), a finding also confirmed by other researchers (Milgate et al., 2002). And our data also suggest that trade union training was more effective than training from other sources (Garcia et al., 2007).

In conclusion, after ten years of the first law allowing specific workers' representation for occupational health and safety in Spain, Spanish safety representatives are mostly men, aged around 40 or so, with a long time working in their companies (around 15 years) and with 'permanent' contracts. Most of the activities developed by Spanish safety representatives relate to providing information and advising workers. Activity is also greater in larger workplaces and in the industrial sector. Although a majority of safety representatives expressed support from other agents involved in risk prevention, supervisors and occupational health services are identified as obstacles with some frequency. Less frequently, some safety representatives find that workers, labour inspection and other trade unions are obstacles too. Satisfaction with support from their own trade union is in general high, training and information being the lowest-rated supports, according to interviewees' opinion. Although the frequency and hours of training given to safety representatives is generally high, they still feel a lack of training as well as a lack of time for proper implementation of their duties. Most frequently reported health risks in the workplace, independently of sector of activity or company size, are related to physical and psychosocial loads. However, the perception of health and safety in the workplace is in general positive. Most safety representatives consider that workers' health is protected in their companies, although safety representatives point out problems related to health and safety bureaucracy, safety levels in small companies, and economic arguments against health and safety investment.

According to our results and available evidence, critical points needing further information regarding safety representatives and their activities include answering questions such as:

- Are young workers less interested in health and safety issues?
- Does under-representation of women and temporary workers as safety representatives affect workers' protection?
- Is safety and health representation less valued than other representation positions in trade unions?
- Do women suffer additional pressure as representatives?
- Why is interest in occupational safety and health lowest in the smaller workplaces?
- Are safety representatives really participating in and positively influencing occupational safety and health decisions in companies?
- Do safety representatives in small workplaces and non-industrial sectors need additional support for proper implementation of their duties?

- Should trade unions increase their focus on sectors or companies covering the largest number of workers in our society (such as services or small companies) and on sectors suffering from major health and safety problems (such as construction or agriculture)?
- Is the kind of training received by safety representatives suitable for the implementation of their duties as workers' representatives?
- How can the time available for safety representatives in the implementation of their duties be more efficiently used?

Notes

We thank safety representatives participating in the research. We also thank technicians from ISTAS and CC.OO. for their advice and help in developing the questionnaire, managing the safety representatives' databases and discussing the methods and results of this study. This research was carried out with the financial support of the Fundación para la Prevención de Riesgos Laborales (Foundation for the Prevention of Occupational Risks).
1. A full Spanish report of the survey is available at www.istas.net. Related scientific papers include Garcia et al. (2005) and Garcia et al. (2007).
2. UGT and CC.OO. are the two main Spanish trade union confederations, covering a large majority of unionised workers in Spain. CC.OO. has been particularly active in relation to occupational health and safety at workplace level.

5
An Afterword on European Union Policy and Practice

Laurent Vogel and David Walters

There are over one million health and safety representatives in the European Union (Menendez et al., 2008). They are the result of a long history of trade union support, campaigns and legislative actions at both national and EU levels. Despite this, there is considerable variation between member states in provision and practice; and systems for representation and consultation on health and safety are far from complete or comprehensive – as earlier chapters demonstrate. In this chapter we reflect on some of the issues that lie behind this variation. We first look at how workers' safety representation has fitted into EU policy developments from the 1970s to the present time. We examine some of the features of national systems for health and safety representation, which demonstrate that a quite substantial part of the European workforce has no such representation at present. We conclude with some comments on the very serious gaps discernible both in EU preventive strategies and in research into the role of safety representatives in a worker-driven momentum for prevention of accidents and ill-health in the workplace.

Workers' safety representation and the development of EU policy

The European Union is the outstanding example of an international community of member states organised in ways to promote both trade and economic growth that at the same time has acknowledged the social dimension of these matters and undertaken to include it in its policies (Szyszcak, 2000). Protection and promotion of health, safety and well-being in the EU labour force are fundamental in this respect, as are measures to ensure social dialogue at workplace, industry and national levels. However, these two approaches – economic development and ensuring the well-being of workers and workers' voice – are not without their contradictions in the increasingly neo-liberal market strategies of the European Union as it struggles to respond to pressures of globalisation. Their consequences are illustrated by the vicissitudes evident in the attitude to worker representation on health and safety within the strategic approaches to health and safety espoused by the European Union in recent decades.

There is clearly a connection between what occurs at the level of the European Union and what happens at national and sector levels. A regulatory trajectory on worker representation in health and safety at work has been discernable in member states of the EU since the early 1970s, and several different approaches are evident. It is also possible to observe a two-way process of influence between member states and the European Union operating upon this trajectory. Outcomes in terms of practice at the workplace level are, however, far more difficult to evaluate.

Workers' health and safety and the legacy of 1960s radicalism

If we look at workers' safety representation from a medium-term historical perspective, it is apparent that the 1970s were a watershed moment on two counts. One was the wave of the most militant and radical direct workers' action over working conditions that swept across almost all Western Europe.[1] From Sweden to Italy (Carnevale, 2007) and from the UK to Germany, the link between health and safety and a broader approach to working conditions typified a welter of trade union initiatives over more than a decade, which began in the mid-1960s, peaked around the mid-1970s and were still showing signs of life in the early 1980s.[2] National developments obviously varied widely, but some common threads can be picked out (Bagnara et al., 1985):

First, a broader approach to working conditions emerged in which health and safety were seen less in technical terms and more as linked to core elements of work organisation (questioning the Taylorist organisation of work, demand for humanisation of work, questioning of the chain of command, and so on). In many cases, this new militancy around working conditions brought on a crisis in traditional prevention practices. One of Italian labour's demands in the 1970s was to do away with company medical services, for example. Occupations specialised in prevention were in varying degrees of turmoil all across Europe.

Second, trade unions often found themselves called into question and forced into a strategic rethink (Hyman and Ferner, 1994). New groups (women, young workers, immigrants) had spikier relations with the trade unions, demanding renewal and more democracy from them. Some countries witnessed a resurgence of forms of workplace representation and organisations which not infrequently ousted the established system of full-time trade union officials as the union's centre of gravity.

Third, a wholesale rejection occurred of the monetisation of risks. The demand was for prevention more than for compensating the victims of work-related accidents and diseases.

Fourth, this company workforce-driven movement exerted strong leverage in all European countries, resulting in sometimes quite far-reaching legislative reforms, including the different Nordic reforms (about Finland and Sweden, see Elling, 1986) the Workers Statute (1970) in Italy, expanding the health and safety committee to include working conditions in France, and the regulations on safety representatives in Great Britain, to name but a few.

Fifth, with the resurgence of mass unemployment brought about by economic crisis, the established gains of previous years found expression in a demand for control over technological changes. It was an issue that seized the agenda with singular force in the late 1970s and early 1980s (ISE, 1982 and 1985; Knudsen, 1995). There were several aspects to it. Defensively, the demand for a negotiated introduction of new technologies was about containing job losses, upskilling, and renegotiating pay categories. Offensively, it was about making business management more democratic, making work organisation something to be discussed rather than imposed by employer's diktat. Industrial democracy was seen as a central objective. In Germany, it was very linked with the concept of a human design of work (Altmann, 1992). This approach chimed with two other political trends. One was the development of a political ecology, bringing a crisis of faith in progress, and the beliefs that technology was neutral and that technological progress brought improvements to society and the living conditions of all its members. There was some overlap between the trade union desire for a negotiated introduction of new technologies and the more general idea of the need for a social and political debate on technology and science. It is no coincidence that the big thinkers most active in health and safety at work during the 1970s pioneered some areas of development of ecological criticism.[3] The same trend can be seen among some trade union leaders. The other development was probably more significant in France, Italy and Spain, where the Communist strand of the labour movement was particularly strong. It resulted from a critical analysis of the Soviet experience and argued that democratisation of work organisation was a necessary part of a socialist society. In part of the Italian trade union movement in particular, the claim for workers' control directly linked an immediate demand to a blueprint for a more-than-capitalist society (Trentin, 2001). Fairly similar debates took place in France's Confédération Française Démocratique du Travail (CFDT) in the early 1970s.

Developments at European level

The contrast between these robust debates and the timorous social policies of the European Community in the early 1970s could not be starker. The social and political crisis that began around 1968 and the downturn in the business cycle of 1973–4 plunged the Community's further development into its first-ever crisis of legitimacy. Hitherto, the dominant belief could be summed up as a virtuous sequence: the creation of a single European market and increased competition would necessarily produce economic growth. This in turn would be sure to stimulate social progress so that the functioning of the Common Market would drive living and working conditions upwards. By the early 1970s, this belief was in some doubt, which helped stimulate the emergence of a new theme around 1974: that the further development of the Community depended on adding a social dimension to the economic one. On the wider agenda were issues concerning, for example, workers' representation in multinational corporations.[4] Another focus of debate was the negotiated introduction of new technologies, where a barrage of employer opposition scuppered a putative directive. Finally, the issue of workers' influence in the decision-making process of technological change was channelled without real result into the 'social dialogue' initiated by Jacques Delors. A specific working party on 'social dialogue and new technologies' was created and met regularly from 1985 to 1988. Legislative instruments were created only in the event of mass redundancies.[5] By definition they related only to potential consequences for jobs, and did not allow for prior negotiation of technological choices. The provisions of the first Collective Redundancies Directive limit the information to the issues directly linked with employment, and do not allow the negotiation of strategic choices about the company policy and technologies.

The fresh impetus health and safety received as a result of the 1974 Social Action Programme focusing on employment protection and industrial democracy in part reflected its inclusion of a specific reference to health and safety. It led to more technically oriented new Directives, for example those on safety signs (1977) and vinyl chloride monomer (1978). It also led to the establishment in 1974 of the Advisory Committee on Safety Hygiene and Health Protection at Work, which became the main forum in which employers, trade unions and representatives of national authorities were to debate the development of detailed policy on health and safety in Europe. In 1978 the first Action Programme on health and safety was announced. At the same time, recession, high unemployment and economic competition from the Far East and the United States prompted a restructuring of industry across Europe. This helped to constrain legislative action, even in relatively non-contentious areas of employment law such as health and safety, to minimal levels. The most significant legislation made under this programme was a framework Directive on the control of chemical, physical and biological agents at work, known as the 'harmful agents Directive' (80/610/EEC, later amended by 88/642/EEC). There were further Directives on lead (1982), asbestos (1983) and noise (1986). A second Action Programme followed in 1984 extending the areas covered by the first. Progress within the Council of Ministers was slow, the requirement for unanimity effectively giving individual member states the opportunity to veto new requirements.

During the 1980s the architects of the Internal Market paid more attention to the importance of its social dimension. This was necessary, it was argued, to offset the potential of the Internal Market to otherwise promote 'social dumping' and to avoid a so-called race to the bottom where social protection was reduced or abandoned in the face of the pressures of economic competition.[6] Despite this, progress with Directives on health and safety remained painfully slow, mainly because of the requirement for unanimous agreement in the Council of Ministers before they could be adopted.

In 1986 the Single European Act heralded a new phase of European-level action. Several aspects of this development are pertinent. Prior to the Act, the Delors 'vision', emphasising the social dimension of the internal market, had also found expression in the promotion of informal strategies for social dialogue between employers' and workers' organisations. They were formalised in Article 118 B of the Act and heralded future agreements between the social partners, such as the relatively recent ones concerning stress and violence at work.

A far more obvious consequence of the Act and one with tangible effects on health and safety legislation was its provision under Article 118A for qualified majority voting and the introduction of the new cooperation procedure with the European Parliament on health and safety Directives. The removal of the requirement for unanimous agreement before health and safety Directives could be adopted effectively ended the ability of some countries to veto their contents.[7] This meant that, under the third Action Programme adopted in 1987, it was possible to introduce an ambitious legislative agenda at European level in time for the completion of the single market by the end of 1992. Spearheading the resulting Directives was the EU Framework Directive 89/391 on the introduction of measures to encourage improvements in the safety and health of workers at work.

Following the Single European Act 1987 was the Charter of the Fundamental Social Rights of Workers in 1989, which was approved by 11 of the then 12 member states of the Community. Although not part of the Act, but basically a political decision that was accepted by a majority of member states, the Action Programme adopted as a result was a further drive towards the implementation of the unprecedented number of health and safety measures under the legal bases of Article 118A. The Framework Directive 89/391 was central to these measures. Its use as a means of establishing a set of 'general principles', which were then followed by a series of daughter Directives dealing with specific risks, has been suggested as a key to understanding this legislative programme (Nielsen and Szyszczak, 1997, pp. 336–44; Neal, 1998).

The Framework Directive and worker representation

With the adoption of the 1989 Framework Directive (89/391) on the introduction of measures to encourage improvements in the safety and health of workers at work, the issue of the representation and consultation of workers on health and safety finally received direct attention in EU requirements.

Under Article 11, employers shall consult workers and/or their representatives and allow them to take part in discussions on all questions relating to safety and health at work. This presupposes:

- the consultation of workers
- the right of workers and their representatives to make proposals
- balanced participation in accordance with national laws and/or practices (11.1)

Article 11 states that workers or workers' representatives with specific responsibility for safety and health shall take part in a balanced way, in accordance with national laws and/or practices, or shall be consulted in advance and in good time by the employer with regard to:

- any measure which may substantially affect safety and health
- the designation of workers involved in preventive services and their activities
- the enlistment, where appropriate, of the competent external preventive services

Such workers and their representatives also have the right to ask the employer to take appropriate measures and to submit proposals to him to that end to mitigate hazards for workers and/or to remove sources of danger (11.3).

The construction of the articles of the Framework Directive makes plain its intentions in requiring an integrated and participative system for the management of health and safety. To appreciate this, its measures on representation and participation need to be seen in conjunction with those on employers' responsibilities to manage risks according to good practice principles of prevention (Article 6), with competent support (Article 7) and through systematic risk assessment (Article 9).

However, the absence of a precise definition of how this regulation of health and safety management is to be achieved is also noticeable. This and the compromises introduced during the debates preceding the final adoption of its measures meant that member states were in effect allowed considerable leeway in their transposition of the Directive.

This was certainly the case in relation to Article 11, for with regard to the practicalities of consultation it left the member states free to set procedures according to national legislation and practices. While the Directive was being negotiated, the European Parliament tried to write a co-decision power into it (Vogel, 1994). But its final wording is much less specific, citing the need for 'balanced participation' – a concept open to very different interpretations, allowing as it does, for example, consultation organised unilaterally by the employer, even on a one-to-one basis – as in the United Kingdom under the Health and Safety (Consultation with Employees) Regulations 1996 in companies where there is no union-appointed safety representative, and likewise in Belgium, in firms with fewer than 50 workers where there is no shop stewards' committee. As Nichols and Walters argue in Chapter 1 of this collection, the effectiveness of such mechanisms is open to question, and they smack of a purely cosmetic transposition, although seemingly enough to satisfy the European Commission.

In other countries, by contrast, transposition of the Directive prompted substantive changes in provisions for worker representation on health and safety. Italy and Spain were particularly interesting cases in point. Prior to its transposition workers' representation in both countries was most often through 'general purpose' mechanisms like works councils or shop stewards' committees. The safety representative concept was a big innovation in both Italy and Spain. The election of tens of thousands of such representatives in each of these countries helped give a renewed impetus to workplace prevention.

In the years following the Directive's adoption and in the very long period during which its implementation in national systems was under way,[8] its status as the cornerstone of a European approach to systematic health and safety management gained considerable international acceptance (see for example Frick et al., 2000; Walters, 2002). Worker representation and consultation was one of the central tenets in this conceptualisation of the regulated systematic management of health and safety and indeed one that distinguished its approach from those of many voluntary systems and OHS management standards internationally (Frick, 2008). Despite this, by the early years of the new millennium the wind of EU policy on this issue seems to have taken a somewhat different direction. The European Commission's first stocktaking report on its implementation did an about-turn on the safety representative issue (European Commission, 2004). With no real explanation, the report evaluates safety representation in very negative terms, reflecting the deregulationist stance of some governments more than any serious analysis of practice. The report states (p. 22) in particular:

> In Member States with a clearly defined culture in co-management, a negative trend has been observed concerning the institutionalised representation of interests. The more clearly defined employment relationships and the higher the number of institutions and committees dealing with occupational safety and health, the less likely workers themselves actively participate in the definition of the prevention policy in the enterprise.

As observed in earlier chapters, there is no robust evidence for this remark. Indeed, most of the available national data points to exactly the opposite conclusion: health and safety committees tend to make prevention more systematic and enable a high degree of worker participation. The Commission's lack of precision about the source for its comments is matched only by the vagueness of the wording – a bald reference to 'French experts' (European Commission, 2004), p. 22).

In 2002 the European Commission adopted a new Community Strategy on health and safety at work – the new title for the Action Programmes for health and safety that had been periodically adopted since 1978. It had little to say about worker representation and consultation – which was somewhat surprising given its major role in the regulated self-regulation as conceptualised by the Framework Directive 89/391. But such back-pedalling is less surprising if viewed in the wider context of the growth of the neo-liberal agenda evident within EU political and economic policies more generally.

Since its adoption, the Framework Directive and other Directives from the same period – all facilitated by the requirements for qualified majority voting in the Single European Act 1986 – have had their detractors, who have complained of over-regulation and of regulatory burdens on the competitiveness of business within the EU. In the 1990s, for example, the so-called Molitor Group, comprised of representatives of employers and some then right-wing governments of EU member states such and the UK and Germany, called for deregulation of the Framework Directive and its daughter Directives, without much in the way of concrete evidence to support their view of the damaging effects of these measures (TUTB, 1995). By the time the 2002–6 Community Strategy was under way, such views had become more firmly entrenched within the governance of the EU and corresponded with wider deregulatory concerns both at EU level and within member states. Adoption of the 2002 Strategy by the Spanish presidency coincided with the joint declaration by Prime Ministers Aznar (Spain), Blair (UK) and Berlusconi (Italy) against a social Europe in 2002, as well as with external pressures from sources such as the Bush Administration in the United States – which was seriously opposed to the REACH Regulation on the safe use of chemicals, for example (Waxman, 2004). Subsequently, under the Dutch presidency in 2004, preoccupation was with the costs of preventive regulation and the desire to 'simplify' or replace it with 'soft' law and voluntary measures (this was the main theme of the Dutch presidency health and safety conference in Amsterdam in 2004[9]). As several critics have noted, evidence cited to support such notions was suspect and never properly validated (Vogel, 2004; Woolfson, 2005). Latterly the ascendancy of the Directorate General for Enterprise within the Commission served to hasten the downgrading of support for both the administration health and safety matters within the Commission and the position of regulatory strategies for health and safety within EU priorities overall. One line that has been adopted in this neo-liberal, market-oriented policy environment is to talk in terms of 'better regulation' rather than openly acknowledge deregulation.

Initiatives launched under this slogan put the machinery for consulting workers' representatives in health and safety issues under serious threat. For example, the obsessive drive to reduce the 'administrative burden' of health and safety legislation helps, among other things, to undermine the employer's obligation to draw up documents that are essential for consulting workers. The Vice-President of the European Commission, Günter Verheugen (2008), himself considers that the information obligations of employers towards the public authority, their own workers and other interested stakeholders should be radically reduced. A quite eccentric paper recently presented by the UK authorities ran along similar lines, arguing that the time spent by firms on health and safety management was just so much red tape that should be cut (Hutton, 2008). It is suggested, for instance, that workplace risk assessments could be simplified in a pragmatic approach utilising standardised checklists. But this fails to acknowledge that such an approach would be likely to both downplay risks and disregard the interplay between physical factors and work organisation (Vogel, 2008). The UK proposals are presented under a flashy cover. They promise 6.6 billions euros saving for the EU economy and a benefit of 255.8 millions euros for the risk-assessment simplification, but the basis on which such calculations have been made is unreliable to say the least.

The Commission strategy for the period 2007–12 was introduced in this climate (European Commission, 2007). Its focus is on linking preventive health and safety to improved economic productivity and financial savings for social security systems. It proposes 'simplification' of administrative and regulatory frameworks within the spirit of EU policy on 'better regulation'. It evinces a further significant loss of interest in workers' representation as a feature of systematic health and safety management. The subject is broached only in a disjointed chapter on changing behaviour, which lists a miscellany of factors. It talks, for example, of stimulating the sectoral social dialogue to ensure that *'workers' representatives are given a greater coordinating role in the systematic management of occupational risks'*. This vague wording heralds no concrete initiative.

It is far to soon to evaluate the impact of the current strategy, but it is difficult to see how its economically oriented, vague and insubstantial proposals will exert any significant influence over the range of concrete challenges facing the improvement of health and safety in the diversity of experience in the enlarged EU – including that of the incomplete and inadequate means of representing workers' interests in the promotion of their health, safety and well-being in European workplaces. Moreover, it would appear that continued jurisdictional wrangling within the European Commission, with the Directorate General for Enterprise looking to exercise yet greater power over the other Directorates General, is likely to lead to further emasculation when it comes to implementation of Community health and safety policy.

Workers' representation in national prevention systems

The development of national prevention systems is governed by specific dynamics that stand relatively independent of the Community strategies (Vogel, 1993; Walters, 2002). This is particularly to be seen in areas where Community harmonisation remains very loose and readily yields to national laws and practices. As already noted, in the case of workers' representation on health and safety sometimes Community strategy and Directives have helped to give a fresh impetus to national initiatives that had stalled for one reason or other. That was certainly the case in Spain and Italy and, to a lesser extent, Ireland and Greece. Elsewhere in western Europe there were already regulatory systems in place, and the changes effected as the result of the Framework Directive 89/391 were very limited. There are various reasons why. In the UK and Belgium, there was certainly a political agenda to minimise the impact of Community Directives in this area. In the UK, under a Conservative government it was a general agenda on the organisation of prevention as a whole. In Belgium, it was more a specific sticking point on workers' representation. Other countries – the Nordic countries are a case in point – seem to have decided that their existing provisions amply met the minimum requirements laid down by the Community Directives. While there was some justification for this, even in these countries developments took place prompted by specific national dynamics. Diminishing public support for the different preventive activities in Sweden certainly affected the operating conditions of both company and area workers' safety representatives. Denmark's new right-wing coalition government changed the law to scrap the requirement of having workers' safety representation in firms employing between five and nine workers. In the UK, the introduction of the Health and Safety (Consultation with Employees) Regulations 1996 served little purpose other than staving off possible infringement proceedings at the European Court of Justice and arguably helped undermine the effective operation of existing provisions that were based on legal rights for trade unionists (James and Walters, 1997).

The debates in central and east European countries were particularly complicated, as several factors were in play. In most of these countries trade unions traditionally looked on health and safety as a technical issue more or less apart from the labour relations system. To a large extent, the trade unions controlled the labour inspectorate, but did not consider that it should act directly on health and safety at work issues. In the worst of cases, the trade unions were party to disciplinary campaigns to lay the blame for work accidents on workers. Another problem stems from the very sharp drop in the number of workplaces with trade union representation following the demise of Communist regimes. Privatisation has been key in this. A third factor is the wariness with which most central and east European trade unions have approached forms of representation connected with worker-elected bodies like works councils. Wide between-country variations notwithstanding, one major trend can be singled out – the dwindling strength of trade unions, with the direct result that very many workplaces are without any form of workers' representation. Transposition of the Framework Directive 89/391 produced changes to the law, but they made only limited difference to this broader picture (Stanzani and Kempa, 2004; Woolfson, 2006).

As Charles Woolfson and his colleagues demonstrate in Chapter 7, recent study of the three Baltic republics suggests that workers' safety representation remains far more an aim than an achievement. Political obstacles are key factors in limiting its further development in post-Communist member states, with only a very small minority of employers in firms with no workers' safety representation being willing to allow their workforce to elect representatives.

In a nutshell, broad trends in workers' safety representation in Europe could be summarised in eight points.

1. There is increased specific representation in health and safety matters in countries such as Spain and Italy, which prior to the implementation of the Framework Directive 89/391 had only a single representation-type system.
2. There are wide variations in forms of specific health and safety representation from one country to the next: representatives elected by the whole workforce, appointed by trade unions, elected by or from among works councils, health and safety committees, and so forth.

3. In almost every case these forms of representation are based on the company as a legal entity, rarely going beyond it to allow site or area representation. The main exceptions are in Sweden and, to a less uniform extent, Spain and Italy, where regional or territorial representatives are able to represent workers across a number of small firms in which they are not employed themselves.

4. In some cases what prevents representation being set up is regulations that set thresholds too high. In Belgium, Bulgaria and France, for example, a minimum of 50 workers in a workplace is necessary to set up a health and safety committee. The effect of such rules is to deny representation to large swathes of workers. Other countries have adopted rules that more closely reflect the realities of the work world. The threshold for appointing safety representatives is five workers in Sweden and six in Italy, rising to ten in Denmark, Hungary and Finland.

5. Non-legal obstacles are also important. There can be a wide gap between the number of firms where legislation allows representation to be set up and those where it actually is. This may be the result of weak union presence and/or employers' hostility to health and safety representation.

6. Limited findings from relatively rare national surveys emphasise the lack of adequate resources (in time, training, expertise, and so forth) available to representation bodies in many firms. This may increase difficulties in establishing such bodies or lead to their demise in a number of workplaces.

7. Generally, the development of workers' representation in health and safety is not seen as a priority in national prevention strategies.

8. The spread of contingent employment has two effects. It tends to deny the most insecure workers any form of representation in health and safety. This is so for temporary agency staff right across Europe. It also tends to undermine the effectiveness of existing bodies in subcontracting or shared worksite situations. This is a major challenge for trade union action to improve working conditions.

No broad trend can be discerned in the crucial matter of whether the share of workers covered by health and safety representatives is rising or falling. This is because there is a considerable absence of data from the majority of EU countries (Mendendez et al., 2008).

The very few countries that do have precise, robust data show mixed trends. As Garcia and her colleagues demonstrate in Chapter 4, in Spain the initial impetus given by the Occupational Risks Prevention Act has produced some positive effects. Coverage of the share of workers with prevention representatives in workplaces where a statutory figure has been set is increasing, as is shown by the figures reported by national working conditions surveys. This remains the case despite the close correlation of the presence of representatives with workplace size, the fragmentation of the productive fabric and decline in the industrial sector (where prevention representatives are more widespread than in the services or the construction industry).

In contrast, in the more mature system in place in the UK, as Walters and Nichols and Walters show in Chapter 1, the trend has been is much more uneven since the 1970s, but the most recent surveys show a marked decline in institutional arrangements involving trade union representation on health and safety during the last decade with a parallel rise in so-called direct methods of consultation.

In other countries information is patchy. According to Menendez et al. (2008), for example, Finland and Denmark have around 35,000 to 38,000 safety representatives in cities having between 2.3 million and 2.8 million workers. Portugal, on the other hand, only has a few hundred safety representatives, spread among some 60 firms. Regional surveys in Italy suggest safety representatives were present in some 70 per cent of firms (Instituto per il Lavoro, 2006). Generally, while there is data on trade union membership for different EU countries (average 26 per cent in 20 European countries in 2002 according to the European Social Survey (Jensen, 2005)), and on workplace representation generally, no such comparable data exists for safety representatives.[10]

A grey area in strategies ... and research

A look at national prevention strategies reveals little policy encouragement to developing workers' safety representatives' activities. Most Community countries have set no specific numerical goal or quality objective for developing systems of representation in health and safety. It is as if they were seen as a paper-shuffling exercise whose precise role is ill-defined and underdeveloped. There is a paradox here: most national strategies start from a conclusion that there is a wide gap between the paper rules and the workplace reality. The same gap can be seen between preventive activities as they are actually carried out and the potentials offered by technical and scientific knowledge. To answer the question 'what works?' would require an examination of what has driven changes in health and safety at work both in society and in workplaces. All the available data point the same way: workers' representation is a key component of that momentum and there are certain enabling preconditions for its activity. It is a finding that obviously goes against the belief that lack of prevention is mainly due to lack of interest on the part of the different groups of workers concerned. It is an approach that sees no real conflict over health at work issues, but rather a consensus undermined by too passive an acceptance by the players involved. A glance at history, however, shows the opposite: that health and safety at work is to do with the perpetuation of social inequalities and that the most productive developments have come about when workers have taken united action around them. This neglect of workers' representation is also in evidence in much of the research into health and safety at work. Not many countries have research programmes in this field. Most national research institutions have big budgets for technological and biomedical research, but remain very reluctant to look into the social dynamics. The insights offered by social sciences are broadly disregarded when not confined to purely psychological investigations focused on individual behaviours. This relative disregard by the research fuels a vicious circle between policymaking and scientific activity.

Notes

1. The timeline in the former Soviet sphere of influence is obviously different and we know of no systematic studies on this aspect of the history of work. The demand for control over working conditions to be exercised by elected, removable workers' representatives flared up whenever the grip of the party-state weakened, as in Hungary in 1956 and in Czechoslovakia in 1968, or Poland both in 1956 and in the 1970s. Within this general setting, specific health and safety issues remain shrouded in obscurity.
2. The course of events was held back in Spain by the pro-Franco dictatorship. It was not until the second half of the 1970s that the 'Italian model' prompted a welter of trade union initiatives in Spain (Carcoba, 2007). This lag produced the counter-intuitive result that the unbroken line between this first generation of trade unionists to see health and safety at work as a priority and today's trade unions is probably stronger than in other European countries.
3. The British Society for Social Responsibility in Science (BSSRS), created in 1969, was a case in point – and many of the key players that shaped trade union approaches to health and safety in the 1970s and onwards were themselves originally part of this action group.
4. This was the purpose of the 'Vredeling Directive' drafted in 1980, which triggered a barrage of opposition from European employers and a threat by the American Chamber of Commerce to call a halt to US investment in Europe (Didry, 2001).
5. The first Collective Redundancies Directive was adopted in 1975.

6. See the European Commission's White Paper 1985, (European Commission, 1985).

7. The UK government, for example, had found this a particularly effective means of reducing the impact of the EU on its domestic regulatory policies. How it came to allow the introduction of these new procedures remains somewhat of a mystery.

8. Implementation in EU 15 countries took most of the 1990s. While some countries were quick to introduce minor changes to existing provisions, others, like Italy, Spain and Germany, used the opportunity for a complete overhaul of their regulatory frameworks for health and safety, and took much longer to comply. Additionally, some countries initially implemented the Framework Directive's provisions incompletely and were forced to make further changes following the threat or reality of intervention by the European Court of Justice (Walters, 2002).

9. See http://www.arbo.nl/news/conferentie.stm. Also SZW (2005).

10. According to the Second European Social Survey, in 2004 trade union membership averaged 22.7 per cent in EU 27 (http://ess.nsd.uib.no/).

Part II Challenges and Strategies for Worker Representation in the Modern World of Work

Part II of this book provides several case studies from countries within and outside the EU, in which the challenges to worker representation presented by modern structures and organisation of work are focused on more explicitly. It examines contemporary research findings on the challenges of precarious employment in Canada, and on the position of workers' representation on health and safety in the neo-liberal market economies of the former Communist Baltic states. With an account of Swedish experience, it considers how trades unions and the state have responded to meeting the challenge of representing workers in small enterprises. It also presents a comparative discussion of trade union approaches to supporting worker representation in health and safety in Australia and the UK and the extent to which these approaches are embraced in union strategies to achieve renewal in these countries. The concluding chapter draws the two parts of the book together with some reflections on the problems facing worker representation in health and safety internationally and strategies to address them. In particular, it reflects on industrial relations and regulatory issues raised, identifying common ground in terms of both the preconditions for effective representation and the difficulties of achieving them, and explores possible ways forward in Europe and elsewhere.

6

Precarious Employment and the Internal Responsibility System: Some Canadian Experiences

Wayne Lewchuk, Marlea Clarke and Alice de Wolff

Worker representatives were formally recognised as agents in regulating workplace health and safety in most Canadian jurisdictions in the late 1970s. This was one component of the transition to an Internal Responsibility System that included mandated Joint Health and Safety Committees, right to know regulations, and the right to refuse dangerous work. Very little has changed in this regulatory framework in the ensuing three decades. The effectiveness of these regulations in improving health and safety was contentious in the 1970s and continues to be debated. Earlier work by Lewchuk et al. (1996) argued that the labour–management environment of individual workplaces influenced the effectiveness of worker representatives and Joint Health and Safety Committees. In particular, the framework was more effective where labour was organised and where management had accepted a philosophy of co-management of the health and safety function. The Canadian economy has experienced significant reorganisation since the 1970s. Canadian companies in general face more intense competition because of trade deals entered into in the 1980s and 1990s. Exports represent a much larger share of GNP. Union density has fallen and changes in legislation make it more difficult to organise workers. Non-standard employment, self-employment and other forms of less permanent employment have all grown in relative importance. This chapter presents new evidence on how these changes are undermining the effectiveness of the Internal Responsibility System in Canada, with a particular focus on workers in precarious employment relationships. Data is drawn from a recent population survey of non-student workers in Ontario conducted by the authors.

Worker health and safety representatives and the Internal Responsibility System in Canada

In most Canadian jurisdictions, the legislative recognition of worker health and safety representatives and the transition to the Internal Responsibility System took place in the late 1970s (Walters, 1983). Much has changed in the Canadian economy in the ensuing three decades. Many Canadian companies face more intense competition because of the 1988 Canada-US Free Trade Agreement and the 1994 North American Free Trade Agreement that opened Canadian markets to Mexican and American companies. With trade liberalisation, Canada's economy has become more export-oriented. At the same time, union density has fallen and changes in legislation and workplace organisation make it more difficult to organise and represent workers. Labour markets have also changed. There has been a growth in non-standard employment and, in particular, self-employment and other forms of precarious employment. This chapter explores how these changes have affected the efficacy of worker health and safety representatives and the Internal Responsibility System, and, in particular, how effective the system is in protecting the health of those in precarious employment relationships.

To understand fully the effectiveness of the Internal Responsibility System in protecting the health of workers in precarious employment relationships, it is important to understand how this system emerged in Canada, the immediate health and safety concerns it was designed to address, and changes to the system since the late 1970s. Prior to the mid-1970s, the emphasis in most jurisdictions in Canada was on protecting workers' health via government regulations enforced by government-appointed factory inspectors, sometimes referred to as the External Responsibility System. Growing dissatisfaction with the ineffectiveness of this system and labour's limited role in both drafting and enforcing health and safety regulations led to calls for more participatory rights for workers. The initial push for change came less from the ranks of senior union officials and more from a group of young activists working within workplaces and supporters working on the fringes of official unions (Storey, 2005). These young activists began making health and safety a priority within the union movement and, supported by political allies in government, succeeded in getting several Canadian jurisdictions to pass new health and safety regulations in the mid-1970s.

The call for change was strongest in unionised and male-dominated sectors such as mining and heavy industry. Most of these workers were employed under standard employment contracts in jobs that were full time, permanent and relatively well paid. Workers were concerned about a range of health and safety issues related to work, but at the forefront were exposures to toxic substances, including sulphur dioxide gas, asbestos, silica, radon gas, and violent accidents in underground mines or in steel mills. The health and safety concerns of those in precarious employment relationships were addressed only where they were similar to the concerns of this largely male, unionised full-time industrial workforce. Workplace hazards such as stress, harassment and employment insecurity were barely on the radar screen at this time.

In Ontario,[1] the passage of Bill 70 (Occupational Health and Safety Act) in 1978 mandated the formation of Joint Health and Safety Committees in most workplaces, required most companies to allow worker health and safety representatives, gave workers more rights to know about the hazards they faced, and gave them the right to refuse dangerous work. Most other provinces adopted similar legislation. Unlike in some countries, both workplace Joint Health and Safety Committees and worker health and safety representatives were mandatory in most workplaces in most Canadian provinces (O'Grady, 2000; Tucker, 2003, 2007). Worker representatives were to be either elected by the workers or selected by their union.

To a significant extent, the Internal Responsibility System became a substitute for the External Responsibility System. Assuming there was sufficient common interest between labour and management in the goal of improving worker health and, assuming they had given workers the tools to participate, the government felt less urgency to regulate directly through the passage of detailed regulations or to enforce existing regulations through workplace inspections. The government preferred a 'hands-off' approach for letting labour and management sort these issues out. Evidence of this can be seen in the decline of government-regulated workplace inspections in Ontario by the mid-1990s to less than one-third of their level in the early 1970s, despite the growth in the economy (Tucker, 2007).

From its inception, the Internal Responsibility System in Canada gave labour limited influence over health and safety matters at the workplace. The product of extensive labour agitation during the early 1970s, it was nonetheless a compromise. The 1970s round of health and safety legislation weakly defined the right to know and limited Joint Health and Safety Committees to 'consultative and advisory' roles. The effectiveness of these instruments, from the perspective of workers, was always contingent on how much pressure labour could apply; and this was almost always a function of how strong their union was (Lewchuk et al., 1996; O'Grady, 2000; Story and Tucker, 2006). On paper, Joint Health and Safety Committees were to have access to information associated with potential hazards and actual accidents, had the right to be present when a government inspection took place, and were to participate in investigations of accidents and work refusals. These Committees were empowered to make recommendations to senior management, but the Internal Responsibility System was not meant to diminish management's authority over issues related to health and safety. Some view this as a dilution of the original intent of the Internal Responsibility System (Parsons, 1988). In reality, the breakthroughs on paper were much more limited in practice when workers tried to assert their rights to a safer workplace. A series of conflicts over health and safety in the 1980s are testament to the ongoing tension between labour and management over the issues of worker safety (Smith, 2000).

The labour movement's interest in harnessing the energies of the health and safety activists who had led the charge in the 1970s and operated at the edges of the formal labour movement, and the activists' frustration with the legislation and the lack of progress on health and safety issues, shaped a second wave of health and safety legislation and regulations (Storey and Tucker, 2006). Federal legislation strengthened the right to know with the creation of the Workplace Hazardous Materials Information System (WHMIS) in 1988. WHMIS provided a standardised labelling system for 'controlled products', the provision of material safety data sheets for over 400,000 substances, and worker education and training programmes on how to deal with hazardous substances. Two years later, changes to Ontario regulations made Joint Health and Safety Committees mandatory at more workplaces, required committee members to be trained, and empowered certified committee members to stop work which was perceived to be dangerous.

More critical to understanding the current state of the regulatory systems was the creation of new bipartite organisations jointly administered by employers and union officials that gradually shifted the focus of health and safety regulation away from the shop floor. In 1987, the Liberal government in Ontario created the Joint Steering Committee on Hazardous Substances, bringing together representatives of labour and management with the goal of rewriting exposure standards. The Workplace Health and Safety Agency, a joint labour–management body to oversee health and safety training in Ontario, was created in 1990. These initiatives had barely begun to function when a new right-wing Conservative government was elected in the province in 1995.

With the goal of reducing the footprint of the state and the influence of organised labour, the Conservatives either abandoned or seriously weakened these initiatives in bipartite regulation. Labour mobilisation limited further scaling back of organised labour's role in regulating health and safety. The government backed away from disbanding the Occupational Health Clinics for Ontario Workers and from reducing the inspectorate by 20 per cent.

By the late 1990s, this brief moment of bipartite regulation was largely over. Although the new government had little interest in making labour a partner of any sort, it did not completely vacate the workplace health and safety arena. Workplace health and safety inspections actually increased during these years, as did the value of fines assessed against employers. Some have argued that the increase in the number of inspections had more to do with inspections becoming less comprehensive and the adoption of a policy of blitzing all firms in industrial parks in a day (Tucker, 2007; Storey and Tucker, 2006). The introduction of the practice of allowing inspectors to deal with work refusals over the phone in 2001 further weakened the role of the government agents in Ontario in supporting workplace efforts to manage health and safety (Storey and Tucker, 2006).

On the surface, the regulatory framework of the Internal Responsibility System was largely unchanged by the shift to a more conservative government in Ontario. However, other changes took place that resulted in greater reliance on employer self-regulation, a lesser role for government intervention and a weaker commitment to worker participation (Story and Tucker, 2006). These changes were the product of both the heightened global competition facing Canadian companies and the ideology of the new government intent on dismantling the mechanisms of government regulation in the economy and society in general. The labour movement found itself unable to respond in the way it did in the 1970s. Storey and Tucker attribute this in part to the impact of changes during the brief period of bipartite regulation that had de-politised the health and safety movement and weakened its rank and file support (Story and Tucker, 2006). Bipartite regulation had meant that health and safety discussions moved outside the workplace between appointed union officials, experts and management delegates. To quote Storey and Tucker:

> To the extent that the successes of the Occupational Health and Safety (OHS) movement crystallized in bipartist institutions and forms of education and training, they reoriented OHS activists and trade union officials away from the rank and file and the union local and towards paid union staff and a centralized OHS education/training model and

delivery service. In these ways OHS was removed from the workplace – figuratively and literally – and plunked down in classrooms where, critics of this evolution charge, the political content of the courses has been replaced by an emphasis on the technical and scientific bases of health and safety.

(Storey and Tucker, 2006, p. 180)

Other Canadian research supports the view that the focus of the Internal Responsibility System was moving away from workers as active participants. Comparing results of questionnaires completed in 1990 and 2001, Geldart et al. (2005) argue that senior managers had become less likely to view worker participation as important in improving safety, and that workers felt that management cooperation on this issue had declined. The same research suggests that management was becoming more active on committees and more likely to attend meetings, and was assigning more senior managers to the process. The authors concluded, 'Management now perceives workers as less (rather than more) important for helping them make decisions, while workers now see their joint involvement in company programs as more (rather than less) of a problem for management' (Geldart et al., 2005, p. 234).

The Internal Responsibility System was adopted in Canada at a unique moment in time. The labour movement was near its post-war peak in terms of influence, and the standard employment relationship was widespread. Workers in a number of sectors of the economy felt sufficiently secure that they were willing to demand changes to protect their health, and a cluster of activists beyond the labour movement supported these demands. However, even at its peak the new system improved conditions only somewhat and only for some workers. Where unions had effectively organised workers or where management was willing to co-manage the health and safety function with workers, the Internal Responsibility System had the potential to reduce injuries (Lewchuk et al., 1996; O'Grady, 2000; Levesque, 1995; Tuohy and Simmard, 1993). Recent research has suggested that, even with a union in place, only some Joint Health and Safety Committees are effective. Committees that focus on a technical scientific mode of operation and cost–benefit trade-off arguments were found to be less effective than what were called 'knowledge activist' committees. Health and safety representatives in activist committees gather their own information on risks, place more emphasis on worker knowledge and are more likely to mobilise co-workers to support demands (Hall et al., 2006).

However, for many workers outside of the organised labour movement or working at firms where management kept a tight grip on management rights, the shift from 'external' protection to 'internal' participation had more limited effects. The small gains in participatory rights came at the cost of a general retreat by the government from its role as regulator. As general economic conditions weakened in the 1980s and the 1990s, and as the labour movement was forced to adopt a more defensive position, the limits of the Internal Responsibility System became more apparent. Against this background, changes currently under way in the Canadian economy are likely to diminish further the capacity of joint workplace committees and worker representatives to protect the health of workers. As the next section will outline, one of the most contentious and far-reaching changes in the labour market is the erosion of standard employment relationships and the growth in less permanent, precarious employment.

The Internal Responsibility System and precarious employment

The concerns of unionised, male workers in standard employment relationships were central to the introduction of the Internal Responsibility System in the late 1970s. As discussed above, even for these workers the new health and safety regulatory system had serious limitations. While this class of workers was somewhat typical of the Canadian labour force in the 1970s, it is much less so today. Today, less than two-thirds of Canadian workers are in standard employment. Various economic and policy developments over the last three decades have altered the context in which workers are seeking to protect their health. One key factor is the liberalisation of Canada's trade regime, which largely began with the free trade agreement with the United States and culminated in the lowering of trade barriers with China and other countries. Trade liberalisation has exposed Canadian companies to more external competition while at the same time making Canadian companies more reliant on export markets. In 1970 less than one-fifth of Canadian GNP was destined for the export market. By 2000 this share had peaked at over 45 per cent of GNP (Statistics Canada, 2007). During the same period union density fell by almost one-quarter from a peak of nearly 40 per cent of the non-agricultural workforce in the mid-1980s to around 30 per cent by 2007 (Commission for Labor Cooperation, 2003; Human Resources and Social Development Canada, 2007). In addition, workplace restructuring, such as subcontracting and outsourcing, combined with legislative changes that have restricted and reduced union rights, has made it more difficult to organise new members and ensure legislative protection is extended to all eligible workers.

Of more relevance to this chapter are the changes in the structure of Canadian labour markets. The proportion of the workforce listed as self-employed has more than doubled since the mid-1970s and now represents over 15 per cent of the workforce (Leacy, 1983; Fudge et al., 2002; Statistics Canada, 2008). Part-time employment has increased and now represents about one-fifth of all employees, double the proportion working part time when the Internal Responsibility System was first introduced (Statistics Canada, 2008). Finally, there has been a dramatic increase in the prevalence of temporary employment, rising from around 14 per cent of the workforce in the late 1980s to 20 per cent by the mid-2000s (Vosko, 2007). This trend is not unique to Canada and, as fewer people are employed full time in permanent employment relationships, researchers have begun asking how non-standard forms of the employment relationship are affecting existing health and safety outcomes and the efficacy of health and safety regulatory frameworks introduced in a different labour-market context. This research suggests that workers in precarious employment relationships are likely to face a number of factors that both increase the risk of injury and illness and make existing regulatory frameworks less effective in protecting their health at work (Walters and Frick, 2000; Quinlan, 2000; Quinlan and Mayhew, 1999).

The characteristics of precarious employment almost certainly increase the risk of injury and illness at work. It is generally accepted that an important predictor of lower injury rates is a stable and experienced workforce (Tuohy and Simmard, 1993; Shannon et al., 1996; O'Grady, 2000). As employment becomes less permanent, tenure and experience with a given employer tends to fall, thereby increasing the risk of injury and illness. Precarious employment can also mean employment at multiple worksites and constantly changing tasks and working environments. Legislation regulating exposure to toxic substances assumes a single employer and does not deal effectively with exposures accumulated at multiple workplaces. Perhaps even more problematic, given the rise of non-permanent and self-employment in Canada, are the exclusions and weaker levels of health and safety protection extended to some workers in precarious employment relationships (Lippel, 2006; Bernstein et al., 2006). In some cases, entire classes of workers such as the self-employed are excluded, or sectors where precarious employment is particularly prevalent, such as agriculture or domestic work, are not covered.

Also critical to understanding the efficacy of the Internal Responsibility System in protecting the health of workers in precarious employment are differences between the labour relations context of those in precarious employment and that of permanent full-time workers. Workers in precarious employment relationships are less likely to be unionised and less likely to have an ongoing relationship with either an employer or a group of co-workers. It has already been argued that the labour relations context plays an important role in shaping the effectiveness of the Internal Responsibility System. Rights are effective only if workers have the power to demand employers respect their rights. Unions, and solidarity among workers, make it possible for workers to exercise their limited rights under the Internal Responsibility System, including the right to refuse dangerous work and the right to take advantage of employee representatives in health and safety matters. Lower rates of unionisation and weaker ongoing links to co-workers make those in precarious employment relationships more vulnerable to retribution for defending the rights granted by legislation. As summarised by O'Grady:

> Without the protection of a grievance system, few workers will be inclined to exercise their statutory right to refuse to perform unsafe work. Similarly, only a small minority of non-union members of health and safety committees will summon inspectors to rectify persistent non-compliance with standards. While near universal unionization was not a presumption of the internal responsibility system, widespread unionization – at least in high incidence sectors – was an unstated premise of that system. Indeed, trying to understand the system of internal responsibility and the role of the right to refuse without recognizing the central importance of unions is like trying to put on a production of Hamlet, but leaving out the ghost. ... For an increasing number of workers – increasing both absolutely and relatively – the unstated premise of the internal responsibility system, i.e., the presence of a union, no longer holds.
>
> (O'Grady, 2000, p. 191)

In a series of papers examining health outcomes of workers in precarious employment relationships, Quinlan (2000) suggested that those in precarious employment relationships:

- were less likely to receive job training and health and safety training;
- lacked job specific knowledge;
- experienced either covert or overt increases in workload;

- lacked knowledge and bargaining power to protect their health; and
- were more likely to face work and family incompatibilities.

                                                            (Quinlan, 2000, p. 182)

In a study of Swedish workers, Aronsson (1999) reported that workers in precarious employment relationships were less likely to be knowledge-able about their work environment, felt they were less likely to receive training and felt it was more difficult for them to be critical at work.

The remainder of this chapter uses data from a recent a study on the health impacts of different types of employment relationship con-ducted by the authors to better understand the overall effectiveness of health and safety legislation in light of employment changes. Three questions are explored.

- Do workers in precarious employment face different work-related health and safety risks?
- Can workers in precarious employment assert their right to know through health and safety training and access to information? And
- Can workers in precarious employment exercise their right to partici-pate in health and safety matters at work?

The project

To answer these questions, data collected through a fixed-response, self-administered questionnaire conducted between September and December of 2005 are used (Lewchuk et al., 2008). The questionnaire measures the physical conditions of work, the characteristics of the employment rela-tionship and health outcomes for workers. The questionnaires were solic-ited from 60 Toronto area census tracts representing 145,109 households. Each household in the selected census tracts received a multilingual post-card inviting all members of the household over the age of 18 who had worked in the previous month to participate. Participants were offered Canadian $10.00 for completing the questionnaire, which they could mail, submit by e-mail, or complete online. Questionnaires were avail-able in English, Chinese and Tamil. Posters with tear-off information sheets were posted in public spaces in the targeted areas to encourage more individuals to participate. Those who completed the questionnaire were asked to distribute additional postcards to people they thought might be interested in completing the questionnaire. Approximately 100 interviews with a random selection of survey participants in precarious employment relationships were conducted.

The analysis in this chapter uses the 1854 surveys received from households in the Greater Toronto Area representing non-full-time students who worked for pay in the previous month. It includes surveys from individuals who described themselves as employed under one of three employment relationship categories, including:

- Less permanent employment relationships (n = 316) defined as employed through a temporary employment agency or on a short-term contract of less than one year. Employment could be either full time or part time, but in either case the relationship is temporary;
- self-employed without employees (n = 167); and
- permanent full-time (n = 1371).

The first two categories represent workers in precarious employment relationships. Those in less permanent relationships view themselves as employees, either full-time or part-time, but all working in a relationship that has a degree of non-permanency. Approximately one-third of this group worked less than 30 hours a week in the previous month. The self-employed did not self-identify as employees. However, it would be incorrect to see them as employers or 'entrepreneurs'. They are a class of low-paid contractors, working on their own, many of whom are actually in disguised employment relationships. Nearly 40 per cent worked less than 30 hours a week in the previous month. Our interest in this chapter is to compare the experiences of full-time and part-time workers in precarious employment with the experiences of permanent full-time employees. Given this focus, the 148 permanent part-time workers in the sample are not part of the analysis.

This chapter's focus is the association between employment relationship type and the effectiveness of the Internal Responsibility System. Three compounding factors were included in the analysis: sex, race and employment sector.[2] Relative to Canada as a whole, the sample is representative of men and women over-represented in racialised visible minorities, and has similar employment sector characteristics to the economy as a whole. There was strong evidence that the employment relationship had an independent effect on many of the variables reported in this chapter, even after these three factors are controlled for. This was less true of sex and race, which had an independent effect in less than one-third of the variables reported below. The remainder of this chapter reports results by employment relationship type, for men and women separately. We do not report findings by race or employment sector in detail, but comment on them where relevant.

Characteristics of the sample by employment relationship and sex

Table 6.1 reports the characteristics of the sample and how individuals with different characteristics were distributed across the three types of employment relationship. There were marginally more women (52.1 per cent) than men (47.9 per cent) in the sample. Of some interest, the percentage of men and women overall in less permanent employment and permanent full-time employment was almost identical to the percentage of men and women in the sample as a whole. Men were over-represented and women were under-represented in the self-employed category. Factoring in race results in a more complex pattern. Whites were over-represented in the self-employed category and under-represented in the less permanent category; however, the experience of white men and that of white women were different. White men were under-represented in less permanent employment and over-represented self-employment. For white women it was the opposite, with white women over-represented in less permanent employment and under-represented in self-employment. The same distinction was not found for men and women from racialised minorities. For this group the distribution of both men and women across the three types of employment relationship was virtually identical to their prevalence in the sample as a whole. These results indicate that, while, overall, men and women work in precarious and permanent relationships in the same proportion as

Table 6.1   Sample characteristics by employment relationship and sex (per cent)

		Less permanent	Self-employed	Permanent employment Full-time	*p*
Sex	Male	47.9	53.9	48.7	.410
	Female	52.1	46.1	51.3	
White	Male	29.8	48.9	33.7	.007
	Female	39.6	44.2	40.2	.776
Under 25 years of age	Male	25.2	12.2	8.0	<.001
	Female	23.8	2.6	8.3	<.001
Age 25–50	Male	58.3	57.8	79.6	<.001
	Female	54.3	66.2	78.0	<.001
Over age 50	Male	16.6	30.0	12.4	<.001
	Female	22.0	31.2	13.7	<.001
University degree	Male	55.6	52.2	64.0	.027
	Female	49.4	54.6	57.2	.189

men and women are represented in the sample, this masks a difference between white men and white women. White women are marginally more likely to report working in a less permanent employment relationship and white men are marginally more likely to be self-employed.

There were differences in the age profile of the three employment relationship categories. Men and women under the age of 25 were over-represented in less permanent employment and under-represented in self-employment and full-time employment. Men and women over the age of 50 were over-represented in self-employment, while men and women between the ages of 25 and 50 were over-represented in permanent full-time employment. However, it is important to point out that over half of those in precarious employment were still between the ages of 25 and 50.

Education buffers men from both forms of precarious employment. Men with a university degree were over-represented in permanent full-time employment and under-represented in less permanent and self-employment. A university degree was less effective in buffering women from precarious employment.

Do workers in precarious employment face different work-related health and safety risks?

Injuries and workplace fatalities are certainly linked to employment in dangerous work and exposure to physical hazards. Similar to research findings reported by Quinlan (2000) and others, our study found that those in precarious employment relationships seem to face different work-related health and safety risks from their counterparts in permanent employment relationships. This is particularly true of men. As Table 6.2 shows, men in precarious employment relationships were more likely to report hazardous working conditions than men working in permanent full-time employment. Men in less permanent relationships were more likely to report working with toxic substances, working in noisy environments and working in uncomfortable temperatures. Men in self-employment were more likely to report working with toxic substances. Women in all employment relationship categories generally reported less frequent exposure to physical hazards than men. The differences between women in precarious employment relationships and those in permanent full-time employment were also smaller. Self-employed women were less likely to report working in uncomfortable temperatures, perhaps because of the large number of home-based self-employed women.

Table 6.2 Exposure to physical hazards by employment relationship and sex (per cent)

| | | Precarious employment | | Permanent employment | |
		Less permanent	Self-employed	Full-time	p
Used toxic substances	Male	29.8	22.2	16.2	<.001
at least ¼ the time	Female	10.4	20.8	12.6	.071
Work in noisy	Male	25.2	16.7	14.4	.005
environments at least ½ the time	Female	15.9	9.1	14.1	.365
Experience	Male	28.5	23.3	21.7	.206
discomfort due to air quality as least ½ the time	Female	23.2	13.0	20.1	.183
Work in uncomfort-	Male	33.8	22.2	20.8	.003
able temperature at least ½ the time	Female	29.9	11.7	25.2	.009
Work in pain at least	Male	26.5	15.6	16.9	.018
½ the days in the previous month	Female	25.6	15.6	17.0	.030

Finally, men and women in less permanent employment relationships were more likely to report working in pain than men or women in permanent full-time employment or the self-employed. These findings raise questions about the ability of workers in less permanent relationships to take time off to recover from the more demanding working conditions they face. Workers in less permanent employment are as likely to report their physical workload is too heavy or their work pace is too fast as permanent full-time workers, but they are almost twice as likely to report working in an awkward position at least half the time.

Multiple regression analysis[3] revealed that most of the associations discussed above continue to hold even after sex, race and employment sector are controlled for. Sex was independently associated only with use of toxic substances, and race was not associated independently with any of the other indicators. This is a powerful observation. It suggests that, even after type of work and key individual characteristics are correcting for, the employment relationship continues to influence exposure to physical hazards.

Table 6.3 examines the relationship between the employment relationship and work-related stress. Men in less permanent employment were as likely to report tension at work as men in permanent employment,

Table 6.3 Exposure to stress and harassment by employment relationship and sex (per cent)

| | | Precarious employment | | Permanent employment | |
		Less permanent	Self-employed	Full-time	p
Tense at work at	Male	41.1	26.7	39.9	.044
least ½ the days in	Female	34.2	34.2	43.4	.042
the previous month					
Harassed at work	Male	35.8	20.2	26.7	.021
in the previous	Female	26.2	11.7	22.1	.039
month					
Multiple employers	Male	37.8	30.0	10.0	<.001
create conflicting	Female	30.5	28.6	11.7	<.001
demands					

and more likely to report exposure to harassment and conflicts due to multiple employers. Women in less permanent employment were less likely to report tension at work, as likely to report being harassed at work, and more likely to report conflicts due to having multiple employers. Self-employed men and women were less likely to report being tense at work or being harassed at work but they were more likely to report conflicting demands due to having multiple employers. Although a longer discussion of the issues surrounding self-employment is beyond the scope of this chapter, it is interesting to note that a number of those in self-employment reported having high levels of control over their work, and this appears to be reflected in measures of workplace stress.[4] This relatively privileged group of workers have found a way to benefit from labour market flexibility and to be buffered from some of the more negative consequences of precarious employment.

Multiple regression analysis revealed that, after sex, race and employment sector are controlled for, the employment relationship continued to have an independent association with all three measures of stress in Table 6.3. Both sex and race had an independent effect on harassment at work. Women and whites were less likely to report being harassed at work. The finding that women were less likely to report being harassed at work, after the form of the employment relationship is corrected for, suggests that workplace harassment is strongly influenced by the power relationships implicit in the employment relationship and that this affects both men and women.

Stress associated with harassment at work and the uncertainty linked to multiple worksites may be a potential health risk for workers in precarious employment. This became particularly apparent in our interviews. The health impacts of chronic stress and an individual's limited ability to address health problems linked to stress underscore the fact that protecting workers' health is about much more than occupational injuries and disease. A number of our study participants described how the insecurity of precarious employment relationships affected their health.

An accountant in his early forties working on short-term contracts reported:

> I don't get enough sleep because I don't know what happens from day to day. But I know I've got chest tightness sometimes from stress. I've got a friend who calls me every day to ask me how I'm doing but I say I'm okay. But you know deep down that you're not. You've got all this worry that you look for jobs and there's nothing there for you. I haven't seen a doctor in a number of years. Because I'm afraid to find out what a doctor might tell me.
>
> #2493, May 2006

A nanny in her fifties who reported being self-employed described her concerns as follows:

> No benefits. No sick days off, no overtime. My husband has a factory job ... but his income isn't enough, I must also have a job. I'm always thinking about money. I don't sleep well, I wake up at 2:00 thinking about money. ... I feel very stressed all the time. I think about money. Mentally, it is hard to deal with. You get so stressed you want to yell.
>
> #5621, June 2006

Quinlan (2000) has argued that the shift to less permanent employment creates new workplace risks associated with a greater sense of workplace disorganisation. Our findings lend some support to this hypothesis. As reported in Table 6.4, men and women in permanent, full-time employment are less likely to change jobs and more likely to work in familiar locations than those in precarious employment. Multiple regression analysis revealed that the employment relationship, after sex, race and employment sector are controlled for, continued to have an independent association with both of the characteristics in Table 6.4. Women were less likely to work in unfamiliar locations than men; however, race had no independent association with these variables.

Table 6.4 Exposure to disorganisation by employment relationship and sex (per cent)

		Precarious employment		Permanent employment	
		Less permanent	Self-employed	Full-time	*p*
		Disorganisation			
Average job lasts	Male	52.3	32.6	10.2	<.001
six months or less	Female	66.3	32.9	8.2	<.001
Work in unfamiliar	Male	15.9	24.4	6.0	<.001
locations weekly	Female	9.8	11.7	4.9	.008

Several workers we interviewed noted the range of problems associated with high levels of workplace disorganisation, such as their lack of knowledge or access to safety equipment, a lack of specific training on workplace hazards, and fragmented levels of authority or supervision. For example, a contract worker in his twenties employed through a sub-contracting arrangement with an internet cable company told us that he frequently worked without adequate safety equipment or knowledge of safety issues. He reported:

I needed ladder hooks to hook onto cables on the poles, they didn't have the hooks for the ladders. And they were supposed to have a safety harness for climbing up the poles and they didn't provide that either. There was a lot of safety equipment they didn't really provide. ... So, I had my ladder up against a cable strand on the poles and one of the cables broke and almost fell. So that was very scary ... It would have been a 30-foot fall.

#5208, June 2006

Overall, the findings reported in Tables 6.2 through 6.4 suggest that both men and women in precarious employment relationships face hazardous working conditions more frequently than men or women in permanent full-time employment. For men, the most serious risks appear to be physical risks, harassment and disorganisation, while women report a high frequency of pain at work and disorganisation. For the self-employed, physical working conditions are comparable to or even better than those reported by permanent full-time employees. They are the most likely to report working in unfamiliar locations but the least likely to report tension at work or being harassed at work.

The next section of the chapter explores the efficacy of the Internal Responsibility System in protecting the health of precarious workers exposed to these higher work-related health and safety risks.

Can workers in precarious employment assert their right to know through health and safety training and access to information?

One of the central pillars of the Internal Responsibility System is the right to know through training and access to information on hazardous substances. Table 6.5 suggests that workers overall are not well informed about the hazards they face and that this is particularly the case for workers in precarious employment. Barely half of the men and women employed in permanent full-time positions reported receiving health and safety training at work. For those in less permanent employment, barely one-third of the men and just one-quarter of the women received health and safety training. The share of the self-employed receiving health and safety training was closer to one in five. Survey respondents were equally poorly informed about the toxic substances used at work. Just over half of the men employed in permanent full-time positions who regularly use toxic substances received information on them. The share of men in precarious employment receiving information on toxic substance was closer to one-third. Just over half of the women employed in permanent full-time positions received information on toxic substances. Self-employed women who reported using toxic substances were the least likely to receive information on the substances they were using.

Table 6.5 Health and safety training by employment relationship and sex (per cent)

| | | Precarious employment | | Permanent employment | |
		Less permanent	Self-employed	Full-time	*p*
Received H&S	Male	37.8	16.9	52.8	<.001
training at work	Female	25.2	20.8	51.5	<.001
Received information	Male	37.8	36.8	55.6	.073
on toxic substances	Female	50.0	25.0	59.1	.041
at work (if working					
with toxic substances)					

Multiple regression analysis revealed that the employment relationship, after sex, race and employment sector are controlled for, had an independent association with both variables in Table 6.5. Neither sex nor race was independently associated with receiving health and safety training or receiving information on toxic substances. With the exception of management and administrative occupations, employment sector was also not associated strongly with the prevalence of health and safety training or being informed about toxic substances.

Can workers in precarious employment exercise their right to participate in health and safety matters at work?

A second pillar of the Internal Responsibility System is the right to participate in health and safety matters at work. The findings from the study suggest that the ability to exercise this right is in doubt for a large number of workers, particularly for men and women in less permanent employment relationships. Given the importance of worker participation for the effective functioning of the system, the limited ability to raise safety concerns or participate in a meaningful way in health and safety discussions certainly raises questions about the effectiveness of the entire system.

As Table 6.6 demonstrates, about one-third of the workers in our study reported that raising a health and safety issue or making a compensation claim would be at least somewhat likely to affect negatively future employment. For men and women in less permanent employment, almost half reported raising a health and safety concern would likely affect future employment with their employer, and over half reported making a health and safety compensation claim would have negative employment effects. An even larger percentage of workers reported that raising employment rights might affect future employment, with both men and women in less permanent employment the most likely to indicate such an action would affect negatively future employment. It is hard to imagine men and women in less permanent employment aggressively exercising their right to raise health and safety concerns, seeking compensation for an injury, or defending employment rights, given the high level of concern they have regarding the implications of such actions. For self-employed men and women, the potential impact of asserting their rights was less of a concern.

The survey findings also indicate significant constraints on workers' views regarding their ability to exercise the right to participate to improve health and safety conditions at work. Less than half of the entire sample

Table 6.6   Right to participate by employment relationship and sex (per cent)

		Precarious employment		Permanent employment	
		Less permanent	Self-employed	Full-time	*p*
Raising H&S concern at	Male	49.7	22.5	30.3	<.001
least somewhat likely	Female	42.7	26.3	24.0	<.001
to affect negatively					
future employment					
Making WSIB* claim at	Male	67.6	27.3	34.6	<.001
least somewhat likely	Female	51.2	20.3	31.0	<.001
to affect negatively					
future employment					
Raising employment	Male	73.5	34.8	45.9	<.001
rights at least somewhat	Female	57.9	26.7	41.9	<.001
likely to affect negatively					
future employment					
Raising H&S will lead	Male	31.8	52.3	43.8	.004
to change at least ½	Female	29.9	51.3	40.1	.004
the time					

* WSIB: Workplace Safety and Insurance Board (Ontario)

reported that raising a health and safety issue would result in change half the time or more, less than a third reported it would lead to change most of the time, and less than one in ten reported it would lead to change all the time. Combined with concerns reported above that raising health and safety issues might affect negatively future employment prospects, this raises doubts about the overall effectiveness of the right to participate in Canada. For those in less permanent employment, the right to participate is even more limited. These workers are more likely to be concerned that raising a health and safety issue will affect future employment, but are less confident that raising a health and safety concern will lead to change. Indeed, less than one in three workers in less permanent employment reported that raising a health and safety issue would lead to change half the time or more, barely one in five reported it would lead to change most of the time, and less than one in 20 reported it would lead to change all the time. The self-employed were the least likely to be concerned about raising health and safety issues and the most confident that raising them would lead to change.

Multiple regression analysis revealed that the employment relationship, after sex, race and employment sector are controlled for, had an independent association with all of the variables in Table 6.6. Unlike in previous tables, both sex and race had an independent association with these variables with the exception of sex and the effectiveness of raising health and safety issues, where men and women were equally likely to report raising such issues was unlikely to lead to change. Women and white workers were less likely to report that raising health and safety concerns, making a compensation claim, or raising employment rights would affect employment negatively. White workers were more likely to report raising a health and safety concern would lead to change. With the exception of management and administrative occupations, which were less likely to report possible negative effects from exercising health and safety rights, employment sector had almost no association with these indicators. These findings confirm that the form of the employment relationship influences the effectiveness of the Internal Responsibility System and that those in less permanent relationships enjoy the least protection. However, even after this effect is accounted for, the system appears to have less potential to resolve health and safety issues for non-white workers and men. The latter is a surprising finding and perhaps reflects the increased exposure to foreign trade of male-dominated sectors of the economy.

Interviews with several study participants revealed the vulnerability of workers in precarious employment relationships. A middle-aged women working through a temporary employment agency and diagnosed with carpal tunnel syndrome by her doctor reported:

> I'm not [wearing the brace] because I'm afraid ... Like if anybody sees me I do cry out in pain because it does hurt but I purposely am not going to wear the brace that I have. And as I said to my Mom, even if I do go to the doctor and it does require something, I'm not going to be able to do it until I'm working full-time anywhere.
>
> #5178 June 2006

A young worker working on a series of short-term contracts at a local beer store reported his reluctance to apply for workers' compensation despite being injured at work:

> I just all of a sudden realized that I had a hernia ... I thought about running it through workers comp, but I'm like, as much as they tell ya that that's not gonna affect your employment, that's gonna affect your employment. ... if I took time off for any claim through workers comp, I just, I just didn't think it was gonna bode well ... Actually I've never really known anyone who's gone through workers comp.

I just don't think it bodes well. I think employers see that. I know they say they don't, but I don't believe that.

#2698, June 2006

Even when injuries occur, those in precarious employment are reluctant to report the injury. The same worker described what happens when someone is injured as follows:

[N]o report was written up, like no, like you're always supposed ta..., but no report, I've hurt myself a few times. Like I whacked my face off a shelf, I could barely see, but no report's written up. I don't know, it's just kinda weird how they ah handle stuff over there.

#2698, June 2006

Another worker on short-term contracts indicated that being injured likely meant the end of employment.

I was what was referred to as a special skill actor. I would actually do the work essentially of what stuntmen would do because they're sporting movies and since I have a sporting background I was able to do that. So there's physical labour, and a good chance I'll get injured on the job, so they hire people like me that have experience and, sort of, if they get injured and they just bring in more people. So we're expendable labour.

#5449, June 2006

Table 6.7 reports findings on levels of support at work. It is generally accepted that the rights associated with the Internal Responsibility System are most effectively exercised when workers have the backing of a union and co-workers. It was noted above that union density rates in Canada have fallen in the last two decades and that legislative changes and competitive pressures have limited unions' capacity to act on behalf of their members. Less than one-fifth of workers in the study reported being union members at all places of employment. Unions are virtually absent in the self-employment sector. Men in less permanent employment were less likely to report having a union than men in permanent full-time employment. The difference between unionisation rates of women in precarious employment and those in permanent full-time employment was substantial.

Unions are still able to help workers even when they are not members. Interestingly, men in less permanent employment and in permanent

Table 6.7 Support to defend rights under the internal responsibility system by employment relationship and sex (per cent)

| | | Precarious employment | | Permanent employment | |
		Less permanent	Self-employed	Full-time	p
Union member all	Male	14.6	1.1	19.6	<.001
workplaces	Female	4.9	1.3	19.4	<.001
Union help at least	Male	22.7	11.4	22.4	.055
½ the time if	Female	14.6	6.6	24.9	<.001
needed					
Help with job	Male	32.5	23.3	42.4	<.001
available	Female	40.9	33.8	40.5	.503

full-time employment were equally likely to report a union would help them. However, in both cases less than one in four survey respondents reported a union was there to help them if they needed it. There is also evidence that male workers in self-employment and to some extent men in less permanent employment were less likely to be able to call on the help of co-workers if needed. This seems to be less the case for women in less permanent employment, who were as likely to report they could get help as women in permanent full-time employment.

Multiple regression analysis revealed that the employment relationship, after sex, race and employment sector are controlled for, had an independent association with being a union member and getting union help. Workers in permanent full-time positions were more likely to be members of a union and receive help from a union if needed. White workers were more likely to report belonging to a union.

While the study did not gather any direct information on the third pillar of the Internal Responsibility System, that is, the right to refuse dangerous work, it is hard to imagine workers exercising this right given concerns that using their rights could compromise future employment prospects and the low union density rates.

Conclusions

The purpose of this chapter was to assess the effectiveness of employee health and safety representatives and the Internal Responsibility System in Canada. We were particularly interested in how the shift to precarious forms of employment was affecting the efficacy of the regulatory system. Those in precarious employment were divided into those in less permanent employment, which included those on short-term contracts and working through employment agencies, and the own-account self-employed. Exposure to hazardous working conditions and working with toxic substances on a regular basis remains a reality for many Canadians in this study. Men were more likely to report exposure to physical workplace hazards than women, and men in precarious employment relationships were more likely to report physical hazards than men in permanent full-time positions. More workers in our sample reported stress-related risks than physical risks. Here, the self-employed were marginally less likely to report tension at work. Disorganisation has the potential to increase exposure to health and safety risks for men and women in less permanent employment and self-employment where jobs commonly last less than six months and workers are required to work in unfamiliar locations on a regular basis.

Thirty years after the introduction of the Internal Responsibility System and the formal recognition of worker representatives, there remain major gaps in the health and safety regulatory system for all Canadian workers and especially those in precarious employment. Less than half of all workers reported receiving health and safety training or information about toxic substances they might be working with. For men in precarious employment, it was closer to one in three who received this training, and barely one in four women in precarious employment reported receiving health and safety training. The ability to participate in health and safety discussions at work also appears to be limited by concerns regarding the impact of such actions on future employment and the erosion of workplace support from unions. Men in less permanent employment were the most likely to report that taking actions to exercise their right to participate in shaping working conditions would affect future employment negatively. Women in less permanent employment were slightly less likely than men to report concerns that exercising rights to participate would affect future employment but still reported such concerns far more frequently than women in permanent full-time employment. Less than one-third of the men and women in less permanent employment reported raising health and safety concerns would actually lead to change on a regular basis. Self-employed men and women generally were less concerned that raising health and safety issues would affect their employment and more likely to report that raising health and safety issues would lead to change.

It is obvious that the Canadian regulatory system is flawed and that changes in the 30 years since its introduction have acted to limit its effectiveness even further. The emerging pattern of weaker protection for those in precarious employment is a concern. There are a number of steps that can be taken to improve working conditions. We agree with the suggestion by Storey and Tucker that there is a pressing need to enhance the level of external regulation and increase the role of the inspectorate (Storey and Tucker, 2006, p. 178). Given the growth in less permanent employment and the decline in union density, it seems unrealistic to expect worker health and safety representatives and Joint Health and Safety Committees, as currently structured, to be effective. Others have argued for changes to how Joint Health and Safety Committees function, including the option of imposing on employers a duty to bargain with these committees in health and safety matters (O'Grady, 2000) or giving them new authority to deal with health and safety issues at work rather than simply advising management (Digby and Riddell, 1986). Ultimately the vulnerability of those in precarious employment needs to be faced if we are to continue to rely on employee voice and participation at work to move the health and safety agenda forward. As cautioned by others, doing so will not be an easy matter (See Walters and Frick, 2000, pp. 55–6; Storey, 2009).

Notes

The authors would like to thank Ashley Robertson, who helped with some of the initial research for this chapter, and Andy King, who helped inspire the original project. We would also like to thank Robert Storey for sharing his extensive knowledge of the evolution of the Internal Responsibility System in Canada and Dale Brown for her help in the final preparation of the text. The Ontario Workplace Safety and Insurance Board and the Lupina Foundation funded the project.

1. In Canada, responsibility for regulating workplace health and safety is largely a provincial matter, although certain classes of workers are regulated by federal legislation.
2. The sample was divided into ten employment sectors: management, administration, science-related occupations, health care, education and the public sector, retail and hospitality, construction, manufacturing, transportation, and communication.
3. Findings from multiple regression analysis are based on a model as follows: variable $x = f$ (employment relationship, sex, race, sector).
4. For a longer discussion of stress and different forms of the employment relationship see Clarke et al. (2007) and Lewchuk et al. (2008).

7
Employee 'Voice' and Working Environment in the New Member States: Translating Policy into Practice in the Baltic States

Charles Woolfson, Dace Calite and Epp Kallaste

It has become conventional academic wisdom that among the key social benefits of European Union enlargement for the New Member States (NMS), in addition to enhanced information and consultation rights and non-discrimination measures, has been an improvement of occupational health and safety law and practice through the legislative transposition of EU Directives (Kohl and Platzer, 2004). While some have questioned the success of the institutional transfer of a European 'social model' in the process of enlargement (Vaughan-Whitehead, 2005; Meardi, 2007), the adoption of new legislative frameworks in the area of working environment is characterised as evidence of the spread of a European 'social dimension' to the NMS.

Health and safety is seen as an important area of EU competence and has been regarded as the 'jewel in the crown' of the EU's social policy achievements (Smismans, 2003, p. 55). In those NMS with absent, or only weak, traditions of employee representation in health and safety, it is unlikely that representation for employees in matters of health and safety would have been mandated in domestic legislation, but for the social requirements in the *acquis communautaire* comprising the body of European Union law. This chapter examines the strength of employee 'voice' in terms of participation in the daily processes and practices of workplace safety and health two years after the first wave of post-Communist countries acceded to the EU in 2004. It argues on the basis of a survey of employers and employees that the continuing overall weakness of social dialogue between employers and employees compromises health and safety participation at workplace level.

Working environment is an area which at least superficially might be regarded as 'naturally consensual' – one in which employers and employees can discuss mutual concerns for the collective welfare (a supposedly classic 'win–win' scenario). Alternatively, a body of literature going back to Nichols and Armstrong's (1973) analysis of the proposed reform of UK health and safety law in the 1970s challenges this view of a 'natural identity of interests', and asserts the need for 'independent' worker representation (see also James and Walters, 1999; Beck and Woolfson, 2000). Both contemporary EU policy discourses and most conventional safety management research claim that working environment is one area in which employee voice can be an important factor in stimulating safety and health improvements (Walters and Frick, 2000, pp. 43–65; Walters et al., 2005; Walters and Nichols, 2007). Whether this hypothesis holds true empirically in the NMS is not directly addressed in this chapter. Rather, a prior but related issue is explored, namely, the *saliency* and strength of participative arrangements in the workplaces, against the background of emerging employment relations with intensified effort demands on employees to comply with the discipline of the marketplace in post-Communist workplaces.

The institutional and enforcement failures of the previous regime to some extent colour contemporary attitudes towards participation in workplace safety and health matters. Under the previous regime, anecdotal evidence suggests that independent employee voice in health and safety management was muted by the controlling hand of the Communist Party and the prioritisation of output targets over employee well-being. However, there were at least tacit workforce pressures to protect worker safety which could be brought to bear on recalcitrant management under the Soviet system, especially in hazardous occupations such as mining, calling for a more balanced appraisal of the period (Beck et al., 2002; Woolfson and Beck, 2003, pp. 243–5). Modern concepts of risk assessment were almost entirely lacking, and specific work-related risks were compensated by additional monetary rewards (hazard payments) rather than being eliminated at source. These considerations underline the scale of the task which faced post-Communist societies in creating a modern and safe working environment. That said, whatever the ultimate 'legacy' of the socialist period might be, it is now nearly two decades since that system was abandoned. The more appropriate historical reference point for analysing the current problems of the working environment in eastern Europe is that of the decade-long trajectory of EU accession and alignment, with its accompanying transposition into national law of key Directives relating to workplace safety and health, including those on employee participation.

The chapter proceeds as follows. First, the methodology of the employer and employee survey in Estonia, Latvia and Lithuania, the

Baltic Working Environment and Labour (BWEL) survey, is outlined. Second, based on respondent perceptions, adverse features of the contemporary working environment are mapped. Third, the saliency of various channels of employee voice in the working environment is assessed. Our argument concludes with a view of current European Commission strategy for occupational health and safety in the light of ongoing problems of 'social dialogue' in post-Communist New Member States such as the Baltic countries.

Methodology

The substantive parts of the chapter present recent comparative survey evidence from all three Baltic States of Estonia, Latvia and Lithuania. The *Baltic Working Environment and Labour* (BWEL) survey comprised a sample of 800 employers and 1200 employees conducted in each of the Baltic States in the second half of 2006 and early 2007. The survey population was based on a representative sample of enterprises selected by NACE codes (the European standard for industry classifications) according to the proportional contribution to national GDP, further stratified by regions, ownership and by company size. Respondents in the selected enterprises were chosen using random sampling method from the complete employee list, including both manual and non-manual employees. All percentage figures in the text have been rounded up or down, as appropriate, in order to avoid narrative clutter with decimal points.

A special feature of the survey is that it was the first linked data-set of both employees and employers in Baltic States exploring, inter alia, issues of voice and representation in safety and health. This makes possible analysis of the correspondence of views of the different parties. The main body of evidence reported is subjective opinion, and the data are subject to the usual vagaries of consistency and comparability in cross-national survey research. An important check on the BWEL survey is provided by data from the European Foundation's *Fourth European Working Conditions Survey* conducted in 2005/6, which includes questions on issues of working environment in the New Member States that joined the EU in 2004 (Parent-Thirion et al., 2007). The BWEL survey also has a historical benchmark. It takes forward key themes of two previous surveys conducted in the Baltic States for the Finnish Ministry of Labour, the *Working Life Barometer in the Baltic Countries* (Antila and Ylöstalo, 1999, 2003). Selected questions were incorporated in the BWEL survey from these surveys, and from an exploratory questionnaire administered in pre-accession Lithuania (Woolfson et al., 2003). In addition, questions were drawn from both the European Foundation's *Third Survey of European Working Conditions* (Paoli and Merllié, 2000) and its *First Candidate Countries Survey on Working Conditions (2001)* questionnaire (European Foundation, 2002).

The post-Communist 'Baltic Tigers'

The three Baltic countries of Estonia (population 1.7m), Latvia (population 2.4m) and Lithuania (population 3.4m) have undergone rapid transformation since the early 1990s. While the first decade of transition was marked by significant loss of productive capacity, since joining the EU these countries until recently experienced high economic growth levels. Although income levels are beginning to converge, differences in average earnings remain significant compared with older EU member states. Moreover, recent high levels of economic growth are the result not only of growth in employment but also of long-term ongoing structural changes which, in turn, have produced changes in the working environment and working conditions. Private companies and small and medium-sized companies have entered the economic arena as the dominant type of enterprise in the Baltic States, while the large-scale enterprises more typical of the Soviet period are now exceptions. The backdrop, therefore, is one of economic restructuring and a reconfiguration of employment relations in line with the discipline of market forces.

While there are significant differences between the three Baltic States, they share a strongly 'open market' orientation, and for analytic purposes may be said to comprise a Baltic regional 'cluster'. All three countries are deemed to have 'successfully completed' the transition from planned to market economies, and in so doing have created a 'business-friendly regulatory environment' (Woolfson, 2006). Thus far, market reconstruction has been a success in the Baltic States if measured by GDP (Latvia's GDP growth, the fastest in the EU, reached 11.3 per cent year on year in the second quarter of 2007, followed by Estonia and Lithuania at over 8 per cent) (GS, 2007). If sustainable (there are increasing doubts as to this), such growth rates could provide the basis for a predicted trend towards convergence. In addition, labour productivity growth has been higher than the EU average for several years. According to the Eurostat, while in 1997 GDP per employee amounted to 36 per cent of the EU25 average in Estonia, by 2005 it was already 59 per cent (Eurostat, 2006). GDP per employee in the Latvia and Lithuania has grown to similar extent (respectively by 14 and 16 percentage points) to the level of 48 per cent and 53 per cent of the EU25 average in 2005. Nevertheless, despite these spectacular figures, the Baltic economies remain generally low-wage and low-productivity economies, and still have the lowest labour productivity in the EU25, outperforming only the 2007 EU entrants Bulgaria and Romania.

Rapid economic development has been accompanied by ongoing changes in working environment, both in the effort demanded from employees and in the conditions under which work is routinely performed. In the next section we examine empirical evidence of changes in the working environment in the Baltic States as perceived by employees themselves.

Work intensification in the new market economies

A key feature of the reconfiguration of the labour process in the workplaces of the new market economies has been the systematic intensification of work during the transition period. The shift from planned economy to a market-driven deployment of labour has eliminated many of the areas of underemployment or disguised employment that previously existed in state-owned enterprises. In terms of health and safety outcomes, it has been argued that new intensified work discipline may result in improvements, as tighter managerial control and more individual responsibility on the part of workers positively affects personal injury rates (Pavlinek, 2002). Another, less sanguine view might be that an increase in working intensity may also produce perceived health and safety detriments for employees, particularly of a psycho-social nature.

Findings from the BWEL survey point to a trend towards work intensification, across a range of indicators. Respondents were asked to compare *intensity/working pace changes during the year preceding* the survey. A majority of respondents, 55 per cent of respondents in Latvia, 54 per cent in Estonia and 53 per cent in Lithuania, reported that their work intensity/working pace had increased 'considerably' or 'slightly' in the previous 12 months (Figure 7.1).

The indicators of increasing intensity of work are necessarily subjective; but further questions allow at least some view of the different forms in which employees experience this. Two-thirds (66 per cent) of Latvian respondents and around three-quarters of Lithuanian (71 per cent) and Estonian (77 per cent) respondents reported that they work at very high speed about 'half' or 'more than half of the time'. In addition, a large majority of respondents in all three Baltic States reported working to very

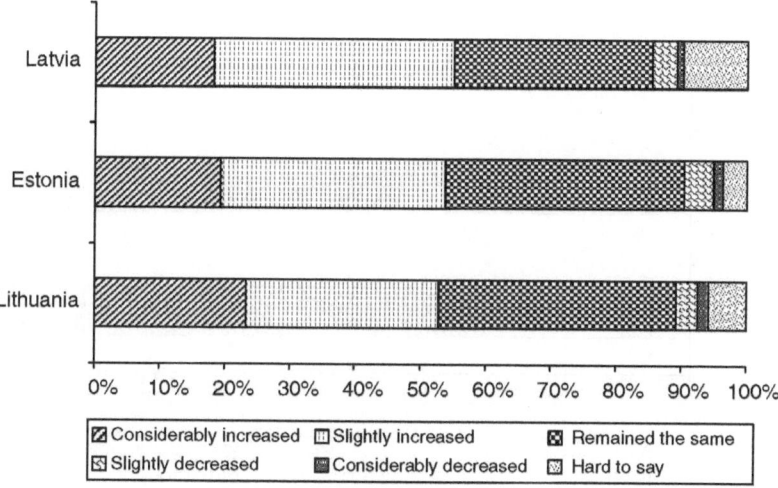

Sample size: Latvia 1,236, Estonia 1,201, Lithuania 1,200

Figure 7.1 Compared with one year ago, how has working pace changed during the last 12 months?

tight deadlines 'half of the time' or 'more than half of the time'; four out of five Latvian (81 per cent) and Estonian (82 per cent) and nearly two-thirds of Lithuanian respondents (62 per cent) reported this. *Physical* intensity of work is still increasing, although it appears to be doing so at a slower rate than previously, perhaps indicating that a physical effort 'ceiling' has been reached. If we compare BWEL data with earlier results from the *Working Life Barometer* (in parentheses), BWEL survey respondents reported that *physical* effort had 'considerably' or 'slightly' increased as follows: Latvia 29 per cent (37 per cent); Estonia 25 per cent (38 per cent); Lithuania 31 per cent (40 per cent) (see Figure 7.2).

When asked to say if working *mental* effort had 'considerably' or 'slightly' increased in the previous 12 months, a larger proportion of BWEL survey respondents than of *Working Life Barometer* respondents reported intensification than previously: in Latvia 46 per cent (40 per cent); Estonia 43 per cent (38 per cent); Lithuania 33 per cent (48 per cent). Only in the case of Lithuania (which had already experienced a sharp increase in reported working mental effort between 1998 and 2002 according to the *Working Life Barometer* surveys), was the reported percentage increase in working mental effort over the previous 12 months lesser (see Figure 7.3).

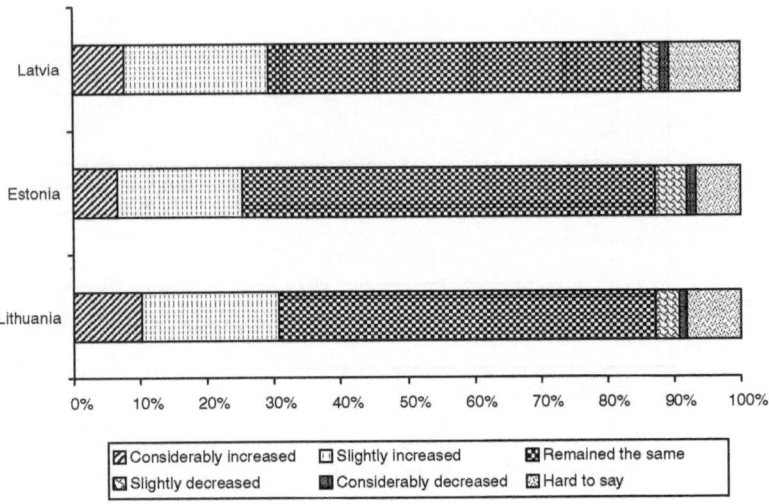

Sample size: Latvia 1,236, Estonia 1,201, Lithuania 1,200

Figure 7.2   Compared with one year ago, how has working physical effort changed?

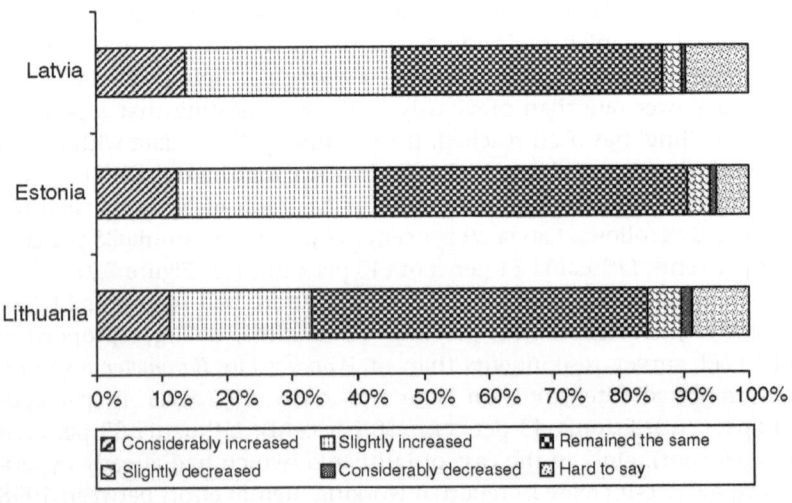

Sample size: Latvia 1,236, Estonia 1,201, Lithuania 1,200

Figure 7.3   Compared with one year ago, how has working mental effort changed?

The European Foundation's *Fourth European Working Conditions Survey* reported increasing work intensity as part of a broader contemporary European picture (Parent-Thirion et al., 2007, p. 58). While there has been a slight decrease in the perception of work intensity in the NMS overall since 2001 according to European Foundation data, this trend was not in evidence in the Baltic States (Parent-Thirion et al., 2007, p. 59). Whatever the underlying factors, employees' perceptions of intensified workplace demands were clearly present. It is helpful therefore to explore in more detail respondents' views of the various most commonly identified physical occupational hazards, as well as those of the psycho-social factors in the working environment.

Working conditions and perceived health impacts

One of the key findings of the *Fourth European Working Conditions Survey* is that a declining number of European workers considered that their health and safety was at risk because of their work, 'although workers in the NMS report significantly higher levels than those in the EU15' (Parent-Thirion et al., 2007, p. 91). The European Foundation data reveal an EU27 average of 35 per cent of workers reporting negative work-related impacts on health, while in the Baltic States between a half (Lithuania 52 per cent and Estonia 59 per cent) and two-thirds (Latvia 64 per cent) claimed this. By contrast, for example, only one-fifth of UK workers did so. In the case of Latvia, the level of reported negative work-related effects was the third-highest figure in the EU27 (European Foundation, 2007). Compared with the European Foundation survey, a somewhat lower proportion of BWEL survey respondents perceived *some* form of risk to their health arising from their work ('certainly' or 'rather') – Latvia (31 per cent) and Estonia (31 per cent), while for Lithuania (40 per cent) the proportion matches almost exactly the European Foundation's reported average for the NMS (see Figure 7.4).

When asked in detail about specific common physical hazards of the workplace, the picture emerges of a poor working environment for many employees in the Baltic States across a range of typical indicators: extremes of temperature, noise or vibrations, poor lighting, lack of adequate working space, exposure to toxic fumes and chemicals, and so on. Overall, these factors were regarded a 'serious' or 'minor' problem by between a quarter and one-third of respondents in the Baltic States, on average more so in Lithuania, and to a slightly lesser degree in Estonia and Latvia (see Figure 7.5).

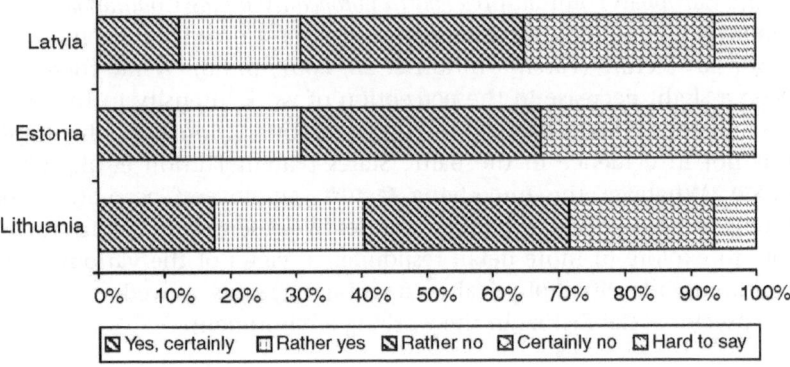

Base: all respondents, Latvia: n=1236; Estonia: n=1201; Lithuania: n=1200

Figure 7.4   Do you think your health and safety is at risk because of your work?

The BWEL survey also posed questions on a variety of psycho-social indicators arising from the work situation which were perceived to impact negatively on health. The particular problems in the three Baltic States reported as 'significant', in rank order, were: fatigue (one-fifth to over a third of respondents), stress (a quarter to over a third), vision problems (one in ten to one in four), anxiety (one in ten to one in four), headaches (one in ten to one in four), irritability (one in six to one in four), and sleeping problems (one in seven to one in four). Overall, Lithuania was the poorest performer on most of these indicators, followed by Latvia and Estonia (see Figure 7.6).

BWEL survey findings are supported by the *European Working Conditions Survey*, according to which stress factors are particularly acute in the Baltic working environment. Negative effects on employees' health resulting from stress were reported on an average more than ten percentage points higher in the Baltic States than in the EU as a whole: Latvia (37 per cent); Lithuania (31 per cent) and Estonia (32 per cent) as against an EU average of 22 per cent (Parent-Thirion et al., 2007, Appendix 3 q33a_q). In summary, the indicators discussed above suggest a problematic working environment in the Baltic States today, as measured by a variety of physical and psycho-social factors, these being perceived as negatively related to health outcomes. The data raise questions, not just about the specificity of the working environment in the Baltic countries, but more generally about an appropriate strategy for securing participatory health and safety improvements at the level of

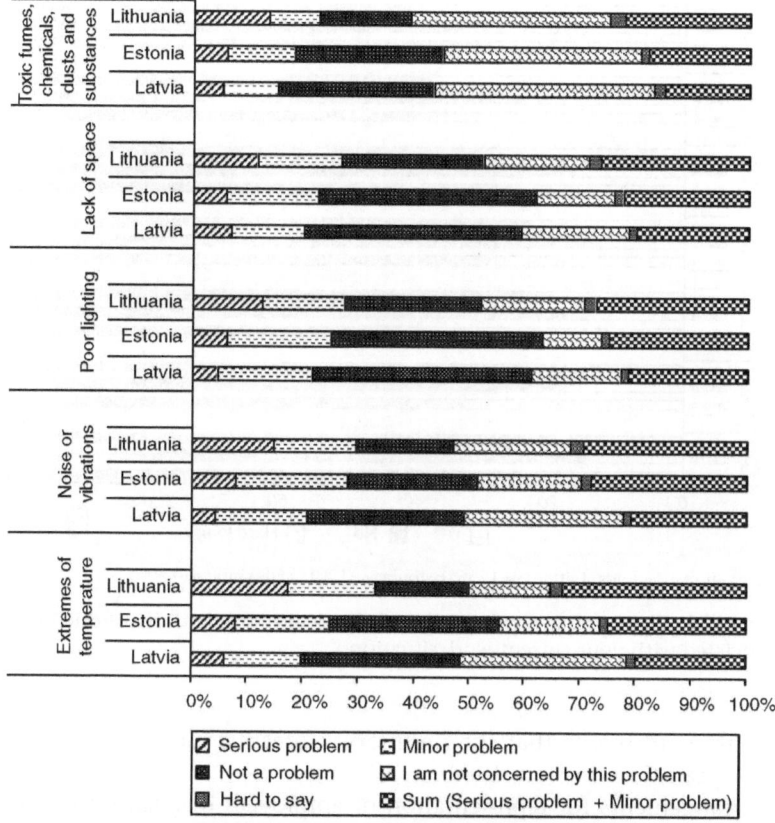

Base: all respondents, Latvia: n=1236; Estonia: n=1201; Lithuania: n=1200

Figure 7.5 Factors which present a problem in your workplace

the workplace. Next, we address the central issue of employee 'voice' in workplace health and safety.

Workforce voice in health and safety

The European Framework Directive on health and safety mandates that 'workers and/or their representatives be informed of the risks to their safety and health, of the measures required to reduce or eliminate these risks', and that 'they must also be in a position to contribute, by means of balanced participation in accordance with national laws and/or

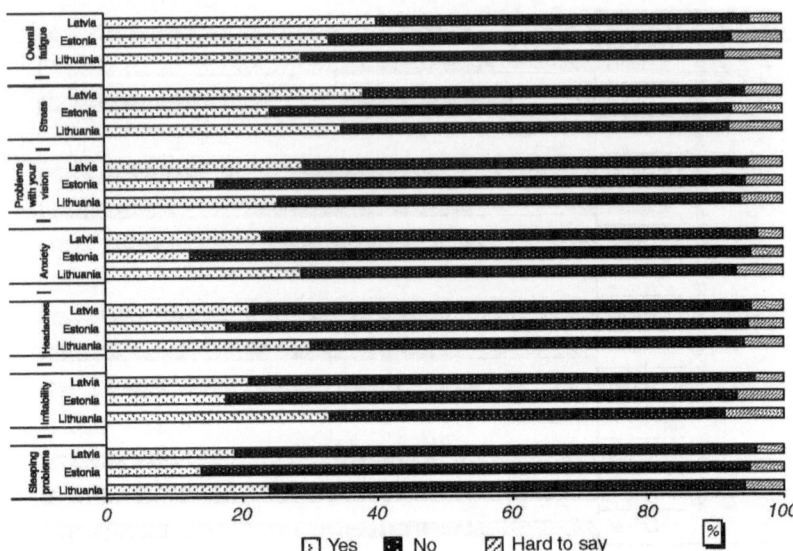

Base: all respondents, Latvia: n=1236; Estonia: n=1201; Lithuania: n=1200

Figure 7.6   Please tell if you have any of the health problems mentioned below associated with your current work situation

practices, to seeing that the necessary protective measures are taken' (European Commission, 1989).

Arrangements for representation of employees in health and safety at work are therefore a general obligation on all member states (part of the *acquis communautaire* comprising the body of European Union law). Each Baltic state, in line with EU requirements, has adopted health and safety legislation in order to promote employee participation in securing health and safety standards. In Estonia the new law on employee information and consultation was recently adopted, and there are also specific laws on health and safety representation and so forth; in Latvia similar legislation has been passed, and in Lithuania both the new Labour Code and a recently amended law on occupational health and safety are in place. This contains extensive provisions for employee participation in safety and health at work (Woolfson and Calite, 2008).

Typically, such representational channels may be either (direct) via individual safety representatives in smaller enterprises, or in larger enterprises (indirect) via elected representatives meeting with employer representatives in health and safety committees. In Lithuania, for example, safety representation is legally mandated in enterprises of 50 or more employees. Levels of compliance with legislation, particularly among smaller enterprises, appear to vary, however (Republic of Lithuania, 2005). In Estonia a similar situation prevails. In Latvia the Labour Protection Law of 1 January 2002 implemented the main provisions of the Framework Directive, although it would appear in real terms that participative arrangements barely exist in most enterprises (International Labour Office 2006). The key question addressed here is how the formal structures of participation are perceived at workplace level by those whose interests they are meant to serve.

One pointer to the *saliency for employees* of health and safety representation arrangements is the degree to which such arrangements are perceived to be in place in the enterprise. The BWEL data also permit comparison between employers and employees on this issue. According to employers in Estonia, a workers' health and safety representative or committee exists in almost all of the companies (97.7 per cent) that have 250 or more employees. In such larger companies, over two-thirds of employees (69 per cent) were aware of these arrangements, although nearly a quarter (23 per cent) responded 'hard to say'. In medium-sized companies in Estonia, most employers (91 per cent) also claimed to have made provision for representation, although only just over half employees (55 per cent) appeared to be aware of its existence. In Lithuania, similar proportions of employers in large (98 per cent) and medium-sized (94 per cent) enterprises claimed to have made provision for safety and health representation. Among Lithuanian employees, however, only 39 per cent in the large enterprises and 35 per cent in medium-sized companies appeared to be aware of these arrangements. In Latvia, again 90 per cent of employers in large companies and 83 per cent in medium-sized companies claimed to have made provision for representation, but only around half (49 per cent) of employees in large enterprises and 43 per cent in medium-sized enterprises appeared to be aware of representation channels. Thus, even where representation is mandatory, there is a clear divergence of view between employers and employees on the extent of their presence. From the BWEL employee data for all three Baltic countries, the picture is not impressive. Overall, among Estonian employees about a third (32 per cent) are aware of the existence of safety representatives or committees; in Latvia somewhat over a quarter (30 per cent) and in Lithuania only a fifth (22 per cent) of employees claimed to know of the existence of safety representatives or health and safety committees in their workplace (Figure 7.7).

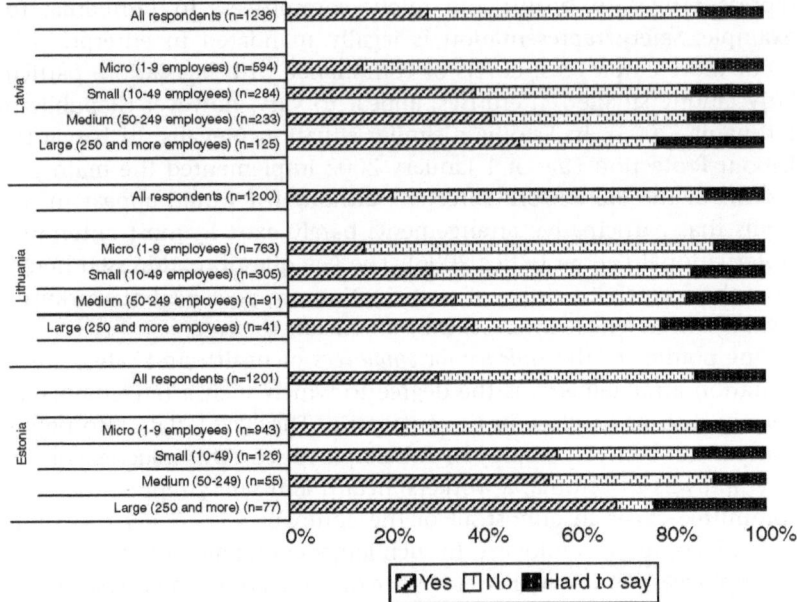

Figure 7.7 Do you have a safety and health representative and/or a health and safety committee member elected from the workforce in your workplace?

Respondents in each Baltic State were asked to identify from which sources they normally receive health and safety advice. In particular, we were interested to know the saliency of elected safety representatives or trade union representatives as active channels of independent voice. In all three Baltic States approximately two-thirds of employees appeared to receive health and safety advice most frequently ('regularly' or 'sometimes') from their supervisor on the job or from fellow workers. Only between 15 per cent (Lithuania) and 23 per cent (Estonia and Latvia) claimed that they 'regularly' or 'sometimes' received information regarding health and safety from safety representatives. An even smaller percentage claimed that trade unions were 'regularly' or 'sometimes' a source of information (Latvia 15 per cent; Estonia 8 per cent; Lithuania 7 per cent). Over a third of employees claimed they 'never' received advice from either of these sources. Put another way, it would appear that the flow of information on safety and health remains largely managerially driven (see Figure 7.8).

BWEL respondents were asked to express their view on desired 'improvements in health and safety' in their workplace. Around half

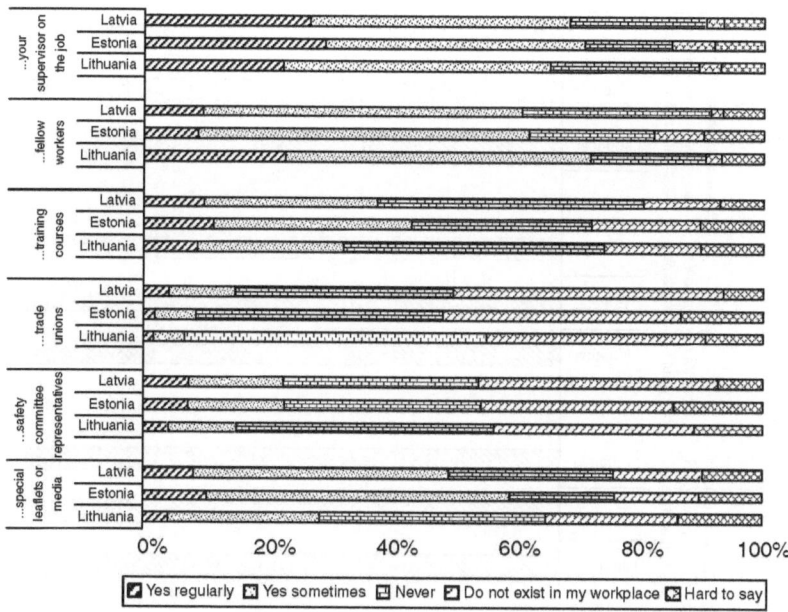

Figure 7.8 Do you normally receive health and safety advice from ... ?

(44–53 per cent) of the respondents expressed the view that there should be 'more co-operation between managers and workers on health and safety issues'. These responses suggest an underlying 'consensualism' on the issue of health and safety – an area where workers and management could jointly consult and contribute. Increased training was a less desired route to safety improvements, supported by only a minority (one-quarter to a third) of respondents. However, the *least* supported safety improvements were 'stronger trade union influence on health and safety questions' (only 16–25 per cent in favour) or 'more power for safety committees to raise issues with management' (only 20–25 per cent in favour). If workers desire to cooperate with management on health and safety, the majority do not see collective forms of representation as necessary to achieving this objective. That said, around one-third to a half of respondents indicated 'hard to say' in response to the latter two questions, indicating a degree of ambivalence or at least reticence on sensitive representational issues. A first reading of the data would suggest therefore a rather depressing outlook, at least from a trade union point of view (see Figure 7.9).

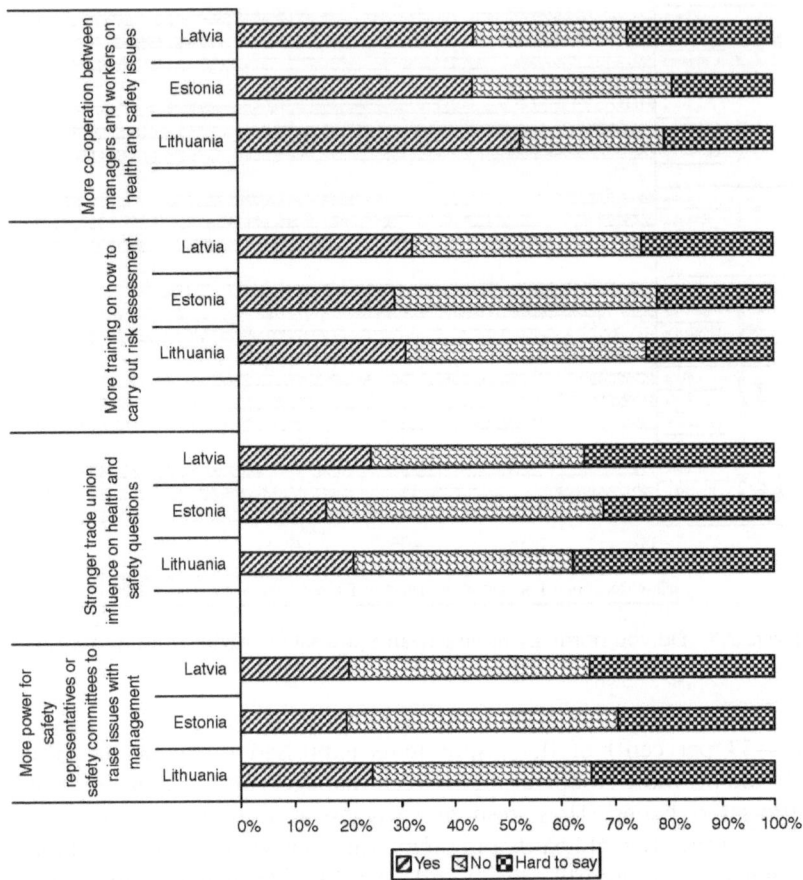

Figure 7.9   What improvements in health and safety would you like to see in your workplace?

Yet, while around a half of all employees appear to favour a consensual approach on health and safety questions, this is not reciprocally reflected in employer attitudes. Employers were asked to choose between possible options for improving employee representation in circumstances where workforce representatives on safety and health had not been put in place. The employer options included 'would consider inviting the workforce to elect such a representative', 'would rather appoint a representative of the workforce to carry out this function who they knew to be reliable', or 'would rather keep things the way they were'. Less than one in ten Latvian employers (9 per cent), one in 20 Lithuanian

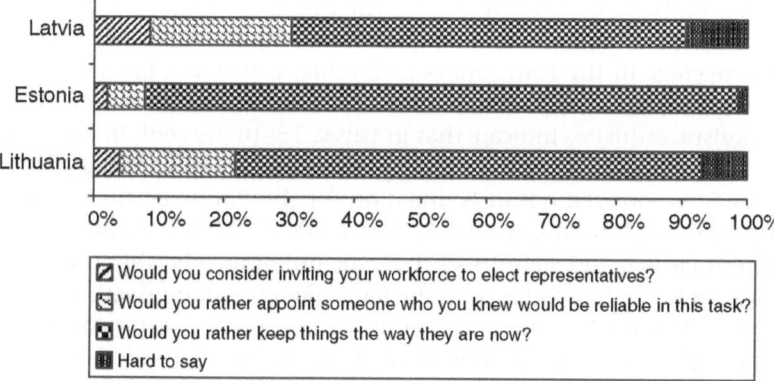

Figure 7.10 If there are no workforce health and safety representatives ...

employers (4 per cent) and one in 50 Estonian employers (2 per cent) would consider inviting the workforce to 'elect a representative to represent the workforce on health and safety'. A minority of employers would 'rather appoint someone who they knew would be reliable': 22 per cent in Latvia, 18 per cent in Lithuania and 6 per cent in Estonia. The overwhelming majority of employers in all three Baltic States would 'rather keep things the way they currently are now': nine out of ten in Estonia (91 per cent), over two-thirds in Lithuania (71 per cent) and over half in Latvia (60 per cent) took this view (see Figure 7.10).

In companies (mainly larger) where safety representatives or committees already exist, these appear to have functional if uneven salience for employees; but where workforce representation is absent, employers have little desire to introduce it, particularly in the smaller and medium-sized and micro enterprises. Here the BWEL survey provides perspectives not replicated by previous studies.

Individualism and collectivism

In western Europe trade unions have often played an important role in the joint governance of workplace health and safety through the system of safety representatives. The evident weakness of collective voice in health and safety representation in the contemporary New Member States would appear to be the corollary of generalised employer hostility and overall employee disempowerment in post-Communist societies. Participation in safety and health by employees cannot be separated from wider issues of participation in the workplace. The role of historical factors in shaping current responses also cannot be ignored. Trade union representation in terms of membership density and collective bargaining coverage in the Baltic States of Estonia, Latvia and Lithuania is in perhaps the least favourable position of all the NMS. Recent somewhat optimistic estimates indicate that in Latvia 15–16 per cent, in Lithuania 14 per cent and in Estonia 11 per cent of the workforce are members of trade unions (European Foundation for the Improvement of Living and Working Conditions (EIRO), 2005a, 2005b, 2006). As elsewhere in eastern Europe, the majority of the trade union membership is concentrated in the public sector or in the few remaining non-privatised enterprises, while union representation in the new private sector companies is negligible and actively resisted by most employers. Where collective agreements exist they are often formal in nature and are 'negotiated' at company rather than sectoral or national level, where labour market actors are weak. In consequence, struggling to maintain a foothold in many enterprises, health and safety representation generally does not feature significantly in the union bargaining agenda.

In order to probe the complexity of employee attitudes towards collective representation, the BWEL survey elicited responses to two sets of questions: the first with regard to trade union issues and the second with regard to health and safety. The responses to the first question set suggest employees are overwhelmingly inclined to seek 'individualist' rather than collective resolution of industrial relations bargaining issues. Four out of five overall in the three Baltic States (80 per cent) agreed that salaries 'are best discussed with the employer on a one-to-one basis'. Only one in five Latvian (20 per cent) and only around one in ten Lithuanian (10 per cent) and Estonian respondents (8 per cent) expressed a desire for the involvement of labour market actors in wage negotiations, either via workplace representatives or national trade unions (see Figure 7.11).

These results mirror the previous *Working Life Barometer* survey findings. The authors of that study concluded, however, that employee responses revealed no more than a realistic appraisal of the current constraints they faced in accessing effective collective bargaining arrangements. In the event that a worker does not have confidence in collective bargaining mechanisms, then 'it may be wisest to rely on one's own negotiation skills' (Antila and Ylöstalo, 2003, p. 80). Thus, 'respondent answers reflect(ed) a subtle contradiction': despite a desire for more collective bargaining, employees 'nevertheless believe that it is wisest to negotiate wages oneself as a general rule' (Antila and Ylöstalo, 2003, p. 80). Some five years later, in the post-EU accession

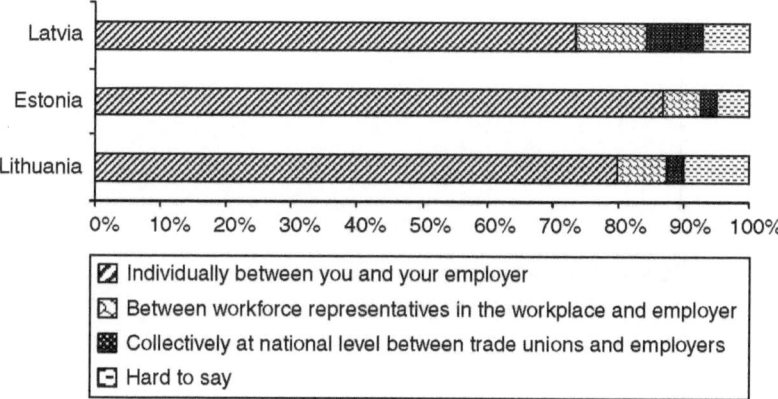

Figure 7.11 Do you think salary issues are best of all discussed ...

Baltic States, BWEL survey data appear to imply that little has changed in this respect.

By contrast, when it comes to issues of representation on workplace health and safety, there appeared to be greater levels of support for more collective approaches. BWEL respondents were asked whether 'employees' safety and health at work issues is best of all discussed individually or collectively'. In contrast to the rejection of collectivist approaches on wages, a substantially higher proportion of respondents were prepared to contemplate a role either for workplace representatives or national trade union actors: in Latvia 45 per cent, in Estonia 31 per cent and in Lithuania 29 per cent. Although still only a quarter to under a half of respondents, it seems that a proportion of employees see working environment as an area where collective forms of representation could be established, although only a small amount of support exists for a role for national labour market actors (see Figure 7.12).

The *Working Life Barometer* survey found a similar contrast between strongly individualist attitudes to wage bargaining and more collectivist approaches towards issues of safety and health. The implications for safety and health appeared to be that 'in the opinion of wage earners, there would clearly be room for more active effort by the trade unions in this respect' (Antila and Ylöstalo, 2003, pp. 86–7). The BWEL data provide further if limited support for the view that if trade unions choose actively to promote working environment issues at enterprise level there may be opportunities to create at least a subsidiary platform for union revitalisation. However, currently trade union capacities in

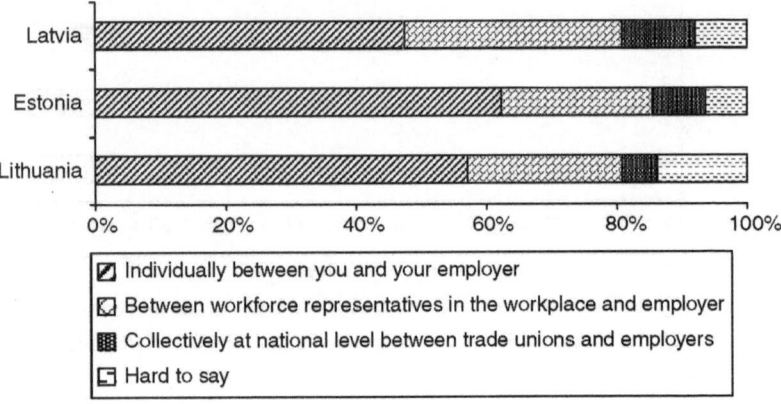

Figure 7.12   Do you feel that employees' safety and health at work issues are best of all discussed individually or collectively?

the Baltic States in this area are almost entirely lacking. In Latvia, where the trade union confederation is in a marginally stronger overall position, a full-time health and safety officer is responsible for conducting free courses on safety and health for union members. In Lithuania, trade union activity on health and safety remains largely confined to tripartite national and regional committees which discuss concerns at a general rather than workplace level. In Estonia, trade union presence in this sphere is negligible at the workplace. Thus, there may be limited scope for trade union activity in which health and safety could offer a 'bridgehead' for renewal.

Conclusion

This chapter has explored employee voice in working environment in the context of an enlarged European Union in which most of the new arrivals are 'post-Communist' countries. The BWEL survey points to significant work intensification, together with defects in the processes of social dialogue. These factors result in muffled representational voice for employees on health and safety matters, despite legislative provisions for employee participation. Currently there is a perception among employees in the Baltic NMS of the low efficacy of collective representational arrangements on both wages and conditions. Employees do appear to be receptive to collective forms of representation on issues of working environment to a somewhat greater degree. The survey data are ambiguous, however, and point to the complex and ambivalent attitudes towards issues of representation and voice in the workplace.

There is a policy dimension to this debate that must be acknowledged. One of the cornerstones of the European Union's previous occupational health and safety strategy (2002–6) has been the attempt to promote a 'culture of risk prevention' in the workplace (European Commission, 2002). This goal was to be achieved by relying 'on legislation, the social dialogue, progressive measures and best practices, corporate social responsibility and economic incentives – and on building partnerships between all the players on the safety and health scene'. Such an approach is predicated upon the encouragement of a 'participative' working environment in which employees can raise issues of workplace health and safety through elected workforce representatives. As a strategy for improving workplace health and safety, it has hinged on 'strengthening the social dialogue at all levels, particularly in firms' (European Commission, 2002). However, the ongoing problems of social dialogue in the NMS remain unacknowledged in current European Commission strategy on occupational health and safety (European Commission, 2007). Outcomes of the EU enlargement process in terms of 'harmonisation' and improvements in workplace safety and health are therefore neither as certain nor as positive as has previously been assumed. Moreover, embracing 'soft law' as the preferred route to safety improvements in the workplace at European level may not be appropriate in the context of the NMS (Woolfson, 2006). Due to the absence of broader social dialogue in the workplace, possibilities for developing collective forms of employee participation, even in relation to a seemingly non-adversarial 'consensual' topic such as working environment, may be limited.

Note

The research was conducted under a European Commission Marie Curie Chair grant (MEXC-CT-2003-509727).

8
Health and Safety Representation in Small Firms: A Swedish Success that is Threatened by Political and Labour Market Changes

Kaj Frick

One of the widely recognised challenges to institutional arrangements for representing workers in workplace health and safety concerns workplace size. In virtually all surveys of the coverage of these arrangements, representation is shown to be increasingly difficult in smaller workplaces. In this respect regional safety representatives have been a success story in Sweden in supporting occupational health and safety in small workplaces since the mid-1970s. Unions may by law appoint them for all workplaces where there is at least one member of a trade union and where there is not already a joint occupational health and safety committee.

Based on the findings of several surveys undertaken in Sweden, this chapter first describes the functioning of the regional representatives scheme and its support, and why it has been successful. It goes on to analyse the extent to which labour market and economic changes in recent years have presented challenges to its operation, and discusses its possible future development in the light of these challenges and their relevance internationally.

A full-scale support system for external worker representation on health and safety in small firms

Regional safety representatives help to improve occupational health and safety in the great majority of small enterprises in Sweden. Although they have resources to visit their workplaces on average only once in a couple of years, this is still many times more frequent than visits to the same firms from labour inspectors or occupational health services, making the regional safety representatives by far the most important actors to provide small firms with competent advice on health and safety. They are able to make an important difference to health and safety in small firms through engaging owner–managers in a dialogue on work hazards.

There is no question that such support is needed in small firms. Workers in these firms face serious risks. Walters (2001) has described health and safety outcomes in small firms within the EU. Swedish firms are no exception. The fatality rate from 1995 to 1998 was proportionally six times higher in firms with 1–49 employees than in larger ones (Antonsson et al., 2002, p. 18). This is not solely because there are more workers in small firms that are involved in manual work or in hazardous sectors, such as construction and transport (SCB, 2003). The main problem across all small firms concerns the contribution to increased risk that results from poor health and safety management arrangements. Studies of health and safety in small firms as well as policy reports on the subject all emphasise the need for external support for health and safety management in these enterprises (Walters, 2001, 2002; Frick, 1979; Frick and Sjögren, 1980; Johansson, 1998; Antonsson et al., 2002; Axelsson, 2002; European Agency, 2002; SOU, 1990, p. 49). Research also demonstrates that face-to-face contact that results from personal visits to small firms and dialogue with managers and workers are most effective, while written information is often ignored (see, for example, Mayhew et al., 1997). At the same time, research and policy discourse within health and safety more widely advocates strong worker participation and representation – in small and large workplaces alike – in order to make this dialogue effective (Walters and Nichols, 2007).

Regional safety representatives combine these strategies. They provide personal support for health and safety in small firms by strengthening workers' representation in management. To mobilise workers in small firms to participate actively in health and safety is acknowledged to be difficult everywhere (Walters, 2001, 2002; SOU, 1972, p. 86; Frick et al., 2005). Various schemes for territorial or roving safety representatives have therefore been set up in different countries, for example in Spain, Italy, Norway, South Africa and the UK. These are generally shown to be effective, both in cooperating with owner–managers and in supporting health and safety improvements. But it is only in Sweden that regional safety representatives have such a developed position and long history of covering the entire labour market (Walters, 2002; Gellerstedt, 2007, p. 11). In countries such as Spain and in Italy the systems are still fairly new, few in number and limited to some regions (UGT, 2008). In Norway, regional safety representatives operate only in the construction industry and only on sites without local safety representatives, while the voluntary approaches in the UK have not gone further than small experiments.

The structure and function of the regional safety representative system

As it is the only full-scale system for external worker representation in health and safety in small firms in the world, the operation and future of Swedish regional safety representatives assume some general importance. Based on a survey, interviews and documents, Frick and Walters (1998) described this system and analysed it within an international industrial relations setting (see also Frick, 1996; Torehov et al., 1996; Frick et al., 1997). In general, the system remains the same today, although, as we shall see, it is increasingly threatened by political and labour market change.

The origins and motivations for regional safety representatives

Requirements on regional safety representatives were first introduced in 1949. Initially limited to small construction workplaces, these were gradually extended to other industries. In 1974 the reformed health and safety statutes extended the scheme to cover all of the labour market. The social partners, in the preceding public inquiry (SOU, 1972, pp. 86, 242), unanimously argued that in small firms:

- there were usually too few employees to find suitable local representatives;
- workers and possible local representatives had too little overview and experience of health and safety issues;
- they were also hard to reach with adequate training and information on health and safety; and
- their employment relationship was often too dependent on the owner–manager for them to be able to participate actively in the health and safety management.

Regional safety representatives were – and are – therefore considered an important support to workers, to possible local representatives and to health and safety management in small firms generally. In contrast to some other schemes (such as in Norway, for example), their purpose is not only to make up for the lack of local representatives and to promote their appointment but also to provide them and their managers with continued support until a joint health and safety committee is set up.

Regulatory requirements

Regional safety representatives may be appointed by regional unions (such as, for example, the Stockholm construction workers union) for all the small firms in their district where they have at least one member and where there is no joint health and safety committee. The large majority of workplaces relying on regional safety representatives are micro firms with fewer than ten employees.

Numbers and coverage

After the 1974 reform and during the 1970s and 1980s, the blue-collar unions within LO (the largest of the three union federations) appointed regional safety representatives for nearly all of their small firms. Increasingly from the 1990s, white-collar unions in the TCO and SACO union federations have also appointed them. In 1995 there were some 1500 regional safety representatives – equivalent to around 320 full-time – who covered 180,000 of the 240,000 relevant small firms, in which some 680,000 blue-collar workers worked (Gellerstedt, 2007, p. 10). By 2006 (AV, 2007) there were in all 1744 regional safety representatives – rather more than in 1995. Of these, 1289 were appointed by LO unions, 393 by TCO unions (that is, by all that organise in small firms) and 62 by two SACO unions. They are nearly all part-time and collectively their work is the equivalent of that of 286 full-time regional safety representatives – down on 1995. They visited some 53,000 of their c. 250,000 small firms (out of some 300,000 which had 1–49 employees) and in these firms there were some 615,000 blue-collar workers in LO unions, that is, amounting to about one-third of all such workers.

Although most regional safety representatives are still from LO unions, the share of TCO and SACO unions has become more numerous and active. This partly reflects a change in the workforce structure towards more white-collar jobs but also a growing interest in health and safety issues among these unions, as manifested, for example, by their health and safety programmes and their appointment of more local safety representatives.

Different types of regional safety representative in different unions

Regional safety representatives – as full-time equivalents – are most common within LO's construction, commercial (such as shops and warehouses) and transport (such as trucking and security firms) workers' unions. The four manufacturing unions also have many regional safety representatives, but they are relatively less common because a smaller share of the membership of these unions works in small firms. Most white-collar regional safety representatives come from the large Unionen – a recent amalgamation within TCO of unions in the manufacturing and trading sectors. But increasingly regional safety representatives are also appointed by public sector unions – for example, by the municipal workers and the teachers – to support occupational health and safety (OHS) in the growing number of small firms created in privatised education, nursing and care. The differences in industries and in unions result in various types of regional safety representative. Especially in manufacturing unions – but also sometimes in service ones – many are chief safety representatives from large workplaces, who take on the role of regional safety representative as a small side task to support health and safety management in neighbouring small firms. In other cases, especially in construction unions but also often in service ones, full-time union officers are appointed as regional safety representatives, also as a side task to their main activity. The transport industry is a special case. Its social partners cooperate on health and safety and training issues, including a partly employer-funded (but still union-appointed) scheme for full-time regional safety representatives. As a result of these variations, the number of small firms covered by individual regional safety representatives in an industry may vary from very few to more than 2000.

Resourcing

By law, the activity of local safety representatives is to be funded by employers. Given forewarning and according to agreements, employers must give representatives paid time off to attend meetings, take part in training, talk to other workers or investigate OHS problems. However, regional safety representatives are temporary visitors to small firms and outsiders. As such it is not feasible to make every small firm pay even a small share of their costs. Instead, since 1974 the unions have been reimbursed for the costs of regional safety representatives through various public means, at present directly through the state budget. The expansion of the representatives' activity since 1974 has not been accompanied by a commensurate increase in public funding. Currently the unions subsidise some 40 per cent of the overall costs of regional safety representatives from their own budgets derived from their membership fees (currently this amounts to some 100 million kronor – about 10 million euros – in state funding and around 65 million kronor – about 6.5 million euros – contributed by the trade unions). TCO unions subsidise a proportionally larger share of the funding but, because they account for the majority of regional representatives, LO unions still contribute much more of their own money overall (AV, 2007). Despite their expanded activity, however, regional safety representatives have only around one-third of the time to support health and safety in small firms that chief safety representatives have in larger workplaces (Frick, 1996).

Role

Regional safety representatives have the same position and rights as local safety representatives; but, in addition, the public inquiry that led to defining these rights also emphasised that they should stimulate local health and safety management, including the appointment and training of local safety representatives (SOU, 1972, p. 86). In practice, because of their size, most small firms have limited internal capacity for good health and safety management arrangements, and the activating role is therefore hard for regional representatives to perform. They therefore mainly function as peripatetic 'local' representatives, who help to detect OHS hazards and discuss with managers how to abate them. Nevertheless, some of their activities can be regarded as increasingly supporting internal health and safety management such as instructing managers and workers how to do their own risk assessments (AV, 2007; see also below).

As to the content of their advice, they support small firms in all types of industry and increasingly also in offices. The health and safety risks they encounter therefore vary widely and so do their interventions to help abate them. Union annual reports in 2006 include descriptions of what regional safety representatives do, covering a range of health and safety risks and including psycho-social issues and health and safety management. For example, in construction they raised ergonomic issues, the use of good personal protective equipment (PPE) and other safety equipment, as well as raising other issues, including safety in scaffolding and electrical work. Occasionally they also stopped imminently dangerous work (while awaiting final decisions on this from labour inspectors). In shops, they discussed stress and ergonomics at checkout counters, including work rotation among tasks, and tried to improve protection against robbery and violence. In private services they achieved a regional collective agreement to improve ergonomics in hotel cleaning, while in small firms in private care they provided training in ergonomic lifting techniques. In manufacturing, they raised ergonomics, safety and chemical issues; and in forestry they informed and trained workers in safe clearing of storm-felled timber. These actions were just some of those undertaken mainly during visits to small firms but also by phone. As in a previous survey of their activities conducted in 1995, the most recent survey, undertaken in 2006, found that their proposals were in the main received positively by managers in small firms (56 per cent) and rarely negatively (7 per cent), with a positive reception more commonly found in services (64 per cent) and a little less so in construction (43 per cent; Gellerstedt, 2007, p. 23).

Regional safety representatives have broadened the scope of their activities compared to those of the mid-1990s (Frick and Walters, 1998). Although they still want more training on work organisation and other psychosocial OHS issues (Gellerstedt, 2007, p. 15), they are now more comfortable about raising such questions (AV, 2007; Frick, 2008). As the above examples show, they proposed work rotation to reduce repetitive work in checkout counters, and acted against stress in offices or in private services, or against risks of threats and violence in everything from security companies to care for the elderly with dementia. In addition, white-collar regional safety representatives have tried to prevent or resolve conflicts in small work groups and have based their activities in this respect in part on implementing the current ordinance against bullying. If this is not possible, they have tried to help affected members find new jobs.

According to union reports, the 2006 survey and a recent meeting with 39 regional safety representatives (Frick, 2008), regional safety representatives also now more actively support local approaches to health and safety management, that is, they carry out the activating role intended for them. Even though small firms have on average become even smaller (SCB, 2008), regional safety representatives nowadays often instruct them how to organise their mandatory arrangements for health and safety management, notably helping them to do risk assessments and providing them with checklists for this purpose. They also try to find new local safety representatives and to support them by training, by phone and by sending them information. For example, 100 new local representatives were elected for restaurants during 2006.

Why it works: The Swedish model of cooperation between social partners

Regional (like territorial and roving) safety representatives are an anomaly within the normal industrial relations system. They are outsiders but still have rights and duties to operate as insiders in small firms, representing their workers. Union officers and other external union representatives regularly interact with local managers (and with their workers and possible local shop stewards). But they do so from an external role, while regional safety representatives are part of the internal dialogue on the small firms' own arrangements for health and safety management. Some of the major factors behind the acceptance of this role in Sweden, and the generally good and successful interaction that regional safety representatives have with managers in small firms, are set out below.

Occupational health and safety issues are legitimate

Raising health and safety issues is usually perceived as legitimate, even in small firms. Especially since the reforms of the 1970s, much public funding has been allocated to produce and disseminate knowledge about health and safety risks and how they can be avoided or abated. Research and development, training, information and occupational health and safety services have spread messages not only concerning the nature and effects of workplace risks to health and safety but also how injuries and diseases can be prevented. The 100,000, mostly trained, safety representatives that exist in Sweden (Gellerstedt, 2007, pp. 47–8) make up a part of this 'enlightenment' strategy, and their presence and context continue to be supported through frequent media reports on risks and injuries, through labour inspection campaigns, and through the distribution of checklists and other material. The latter is also often adapted to the needs of small firms (see for example the website www.prevent.se). While there still is a widespread ignorance on health and safety especially in small firms, managers (and others) who want health and safety information can find it.

The acceptance of the combined messages – that poor health and safety arrangements injure workers but that much can be done to prevent this – can be contrasted, for example, with the situation in the United States, where Nelkin (1985, p. 19) has described serious conflicts at all levels of defining risks and possible actions around workplace health and safety. The recent rejections of federal and state ergonomics standards in the United States demonstrate that this is still the case, and it remains possible to deny the existence of serious risks in the US (AFL-CIO, 2008). This is less likely in the Swedish context where, although there is no consensus on all health and safety issues, even investment bankers share the norm that work health is important and that managers should promote this (Almquist and Henningsson, 2009). Of course, this implies, not that all of the suggestions made by regional safety representatives are accepted in small firms, but that the wider social norms of the environment in which they are made support the legitimacy of making them.

Dialogue within the Swedish industrial relations model

The Swedish (and Nordic) model of labour law and industrial relations, including the law on occupational health and safety, supports the right of regional safety representatives not only to raise health and safety issues but also to get answers. Unions and employer organisations regulate most conditions of work through broad collective agreements. The agreements also cover many issues that in other countries are specified as individual rights in labour law (Bruun et al., 1992). Managers and union representatives – usually local shop stewards – normally interact on much more than wage negotiations. The social partners jointly regard such co-determination as essential in local health and safety management (Frick, 2002). They have also regularly cooperated in national and local development programmes, including on health and safety issues (Frick, 1995; Gustavsen, 2007). When regional safety representatives ask for a dialogue with managers in small firms, they are thus operating as a normal part of this tradition. Despite the weakening of this corporatist tradition since the early 1990s (Frick, 2002) and the further impact on it of recent labour market changes, regional safety representatives still find themselves largely accepted by managers in small firms (Gellerstedt, 2007, pp. 22–4).

Judgement and competence to achieve an insider dialogue

However, a dialogue on health and safety is not enough. Regional safety representatives also have to convince managers of small firms to implement their proposals. In most cases this is helped by their long experience of interacting with managers. Normally, they are familiar not only with health and safety issues but also with the business aspects of their small firms. As such, they can discuss firms as insiders, which enhances managers' acceptance of their proposals (Frick, 1979). With more than 50,000 visits to small firms made by regional safety representatives annually, there are inevitably some conflicting opinions, of which some have to be resolved by the labour inspectorate (SOU, 2007, p. 43, 60–1). Yet, as the figures cited above demonstrate, the advice of regional safety representatives is far more often appreciated than criticised by managers, who also frequently turn to them for advice (Gellerstedt, 2007, pp. 22–4).

Formal rights and leverage over employers

When managers are reluctant to accept their suggestions, regional safety representatives can wield some power to persuade them. With their superior knowledge and practical experience of health and safety, they can often suggest simple and cheap measures – and sometimes even profitable ones – which still can help improve health and safety. They can also often cite regulations in support of their proposals. Although they have no formal enforcement mandate (except for the temporary right to stop imminently dangerous work), health and safety norms make most managers 'voluntarily' accept legal requirements. If nothing else helps, regional safety representatives can threaten to call in the labour inspector; and inspectors are obliged to investigate quickly any complaints that are formally lodged by safety representatives. The formal right to take part in local arrangements for preventive health and safety is thus supported by law, which can be invoked by regional safety representatives against recalcitrant managers.

However, in the end regional safety representatives frequently face a delicate balance between demanding abatement of risks and developing a good dialogue with the managers in small firms. Although they are supported in this by cooperative industrial relations traditions, health and safety norms and formal rights, the competence and judgement of individual regional safety representatives is what determines a satisfactory outcome; and so far they have, by and large, been successful in this.

The basis of the system erodes and difficulties mount

Although the health and safety system of which regional safety representatives are a part still functions largely as intended, changes in the Swedish economy and its labour market have begun to erode its basis. The current conservative government, elected in September 2006, has also reduced support for other actors involved with health and safety in small firms. At the same time broader health and safety issues make workplace visits more complex and time-consuming for regional safety representatives.

A reduced OHS infrastructure with less support from other actors for small firms

Changes in the OHS infrastructure and the labour market have created a need for more support from regional safety representatives for health and safety in small firms. The current government has cut the budget of the Work Environment Authority by some 30 per cent, which will result in 25 per cent fewer workplace inspections (AV, 2008a). It also provides less support for the Swedish 'enlightenment' on occupational health and safety generally. It closed the National Institute of Working Life in 2007 and abolished subsidies for health and safety education and information. The preventive services for occupational health and safety have been shown to have only limited involvement in promoting OHS prevention since the early 1990s (Frick et al., 2005), and their support is particularly limited in the case of small firms (Antonsson and Schmidt, 2003). Finally, unions and employer organisations currently employ fewer health and safety experts than previously. At the same time the number of small firms has increased. Therefore, less health and safety support from a range of sources is provided on a face-to-face basis for increasing numbers of small firms, placing a greater burden on regional safety representatives.

Eroding basis and support

At the same time, there is less employer and political support for the Swedish industrial relations model of dialogue between the social partners to jointly set standards of work and to develop production and working conditions. The proportion of the labour force working for foreign-owned companies in Sweden increased fivefold from 2.5 per cent in 1980 to 12 per cent in 2004 (IVA/NUTEK, 2006). The presence of more adversarial management cultures in these international firms may help to explain why the proportion of managers who believe that local cooperation is good for business fell from 70 per cent in 1996 to 46 per cent in 2003 (Levinson, 2004). The current government has claimed that it prefers the Swedish model of collective agreements between the social partners to more rigid regulation of labour conditions by law (Littorin, 2008). Yet, as we shall see, it has increased the costs of union membership by 30–55 per cent and made jobs less secure. It has also tried to reduce excessive administrative burdens on firms by various means including a review of all health and safety regulations. With higher union fees and labour market changes, union membership dropped from over 82 per cent in 2002 to 73 per cent in 2007, and it continues to fall (LO-tidningen, 2008a).

Fewer members mean poorer unions, with less money to subsidise the activities of regional safety representatives. At the same time, the number of small firms increased by 23 per cent from 1995 to 2005 (SCB, 2008). There are thus many more – and even smaller – workplaces in need of their support (although many more have no union member, and are therefore not covered by regional safety representatives). According to some regional safety representatives, de-unionisation has spread as an active employer tactic to keep them out, for example, among small cleaning firm (Frick, 2008). And although small local firms generally accept them, a level of resistance to regional safety representatives continues to be maintained by peak employer organisations, which are especially opposed to a proposal to counter this erosion (SOU, 2007, p. 43, ch. 2) by giving regional safety representatives right of entry to small firms that lack union members but are nevertheless covered by collective agreements.

The most recent public inquiry on general health and safety at work (SOU, 1990, p. 49) emphasised that OHS could improve only if local health and safety management were more systematic and preventive, and if its coverage included organisational causes and psycho-social risks. As mentioned, regional safety representatives now raise such issues more often than in the past. This is partly because of the increased activity of white-collar regional safety representatives but also because of a generally broader focus on these matters in Swedish health and safety politics since around 2000 (AV, 2006). But organisational issues are usually more complex to address than technical risks. For example, to get a better work rotation roster when managers initially don't understand the organisational health aspects may take several visits. Such issues may also be sensitive. Like proposals on how to improve their health and safety management arrangements, managers may find them unacceptable interference with how they run their businesses. And an individual worker who raises an organisational complaint may on that account lose her or his job (see the case of the cleaner below). Finally, although there are some ordinances on organisational health risks, regional safety representatives have limited legal support when they ask managers to prevent or abate these risks. Their broadened scope, therefore, may well be in line with their original mandate to promote local health and safety management, but it also requires more competence and time from them.

Labour market and political changes weaken the position of workers

The changes that erode the basis for the regional safety representatives system thus simultaneously increase small firms' need for their support. This is accentuated by the general balance of power in the labour market, which has tipped in favour of employers, and especially so in small firms. Workers in these firms have always had a weaker position than those in larger workplaces. Even in the 1970s, when labour had a stronger position – as Nilsson and Carlsson (summary in English in Persson, 1979) confirmed in 1979, a view shared by the inquiry that led to the introduction of the regional safety representative system

(SOU, 1972) – workers in a small workplaces demonstrated that to fight actively for improvements was rarely an option. Although many employees liked working in small firms, their choices were limited, and when they didn't like a job or its health and safety aspects they either had to accept them or leave. In a small firm the social construction of its reality is mainly dominated by the boss, as there is no alternative construction informed by a trade union or other form of workers' collective present (Lysgaard, 1967). Workers in these small firms therefore had a lower rates of sickness or absenteeism ('love it'), but a higher turnover ('or leave it') than those in larger workplaces (Nilsson and Carlsson, 1979). The turnover remains higher (SCB, 2003, p. 23), but recent changes in the economic structure and in the labour market, including the increased role of migrant labour and changes in recent labour politics, have further increased the need for external health and safety support – for both workers and managers.

Most of these challenges are encountered at the lower end of the small-firm sector, that is, in relation to firms that try – or are forced by market pressures – to make profits through paying very low wages and failing to invest in health and safety management arrangements. This experience varies considerably according to the circumstances of different industries. For example, although private services in general have weaker health and safety infrastructures, the sector also includes many profitable small firms with highly qualified employees, whom managers have strong economic motives to care for. Despite the very real problems currently experienced in small firms, there are still large groups of such firms in which industrial relations generally are good and where managers try to prevent risks to health and safety (SCB, 2003).

A globalised and flexible economy and labour market with less secure jobs

Changes in the economic structure and in firms themselves have accelerated since the economic crisis of the early 1990s. Several of these have weakened the position of labour in general, including small firms (as demonstrated in Wikman, forthcoming). The very high unemployment of the early 1990s was down to around 6 per cent in 2008, but this was still much higher than before the crisis. And, in line with increasing pressures for labour flexibility in the market, the threat to workers of unemployment has increased in several ways.

1. Job exports are common. The export–import share of GNP grew from 28 per cent (of both exports and imports) in 1993 to 54 per cent

(exports) and 44 per cent (imports) in 2004. Swedish transnational corporations had about the same number of employees at home as in other countries in 1998, but in 2005 they had twice as many abroad. And, as mentioned, those working for foreign employers increased fivefold between 1980 and 2004, that is, in corporations where jobs are easily moved across national borders.

2. Downsizing has become an important business instrument. For a long time there was a balance between business profits and employment in private manufacturing. However, from around the time of the 1993–4 economic recovery this connection has been lost. Since then, the number of employees in manufacturing has slightly fallen while profits have soared. Layoffs are also common in profitable factories. Higher economic output with fewer workers is a major effect of the continuous increase in productivity in Swedish industry.

3. Existing jobs are more insecure. Temporary employment increased from 400,000 in 1994 to close to 700,000 in 2004 (Wikman, forthcoming) – equal to 16.8 per cent of all employed. Many of them work in the public sector (AFA, 2008) but, for example, 30 per cent or more of private sector hospitality workers are also casual labour (LO-tidningen, 2006a). The rate of casual work is higher among women, among youth, among blue-collar workers and among those born outside Sweden. For example, in the age-group 20–24 years, some 50 per cent work on temporary contracts. It is also higher (14 per cent) in small firms (with fewer than 50 employees) than in larger ones (9 per cent; SCB, 2003, pp. 14, 24). And such jobs are not a means to get a permanent contract. Less than half of the temps obtained such a contract within two years. Labour-hire firms offer another form of casual work. These firms were legalised in 1994 and had some 48,000 employees in 2007, with especially fast growth since 2004. For example, the large engineering industry recently engaged as many labour-hire workers as it had downsized its own personnel (LO-tidningen, 2006b). Franchising has also grown sharply. Of such firms, 68 per cent were set up during the 1990s or later. They now employ some 100,000 persons, who also have insecure jobs, in which conditions of production and work are mainly dictated by franchising contracts.

New job politics further increases job insecurity

Motivated by its main agenda – to bring as many as possible into work – the present government has recently taken several measures to make it easier to hire labour and to make it more profitable to work. However, several of these changes will also serve to increase job insecurity.

1. More temporary employment has been permitted since 2007. As a result employers can now hire personnel on two-year temporary job contracts. With other changes in job security legislation, it is now possible to work for up to four years as a temporary employee.
2. A government inquiry may simplify self-employment by permitting this even when the only customer is a previous employer (Direktiv, 2007). This will allow employers to lay off workers and rehire them as self-employed workers who are owed no employment contract obligations and only very limited health and safety obligations.
3. Unemployment has become costlier. The government raised the fees for union-administered unemployment insurance by 100–300 per cent, especially for LO unions with high unemployment rates, such as in construction and private services. As a result some 10 per cent of the employees have since left the insurance scheme (LO-tidningen, 2008b). Insurance benefits have also been reduced, with lower compensation payments available for shorter time periods. There are tougher demands on the insured to take any job offered. Finally, retraining has become less of an option for the unemployed, as the funding for it has been reduced. Together with tougher sickness insurance policies, this has forced people out of income-related national insurance, leaving them to live at the subsistence level of the local communities' social security (SKL, 2008).

Many more small firms, especially in private services, with few health and safety resources

Less secure jobs have been accompanied by the previously mentioned 23 per cent growth of small firms (with employees) between 1995 and 2005. If small firms with no employees (that is, the self-employed) are included, the number of such firms grew by 57 per cent from 1996 to 2006. Behind this lies the fracturing of the economy, through outsourcing from private companies and privatisation of public organisations. For example, the number of construction firms increased by 22 per cent between 1990 and 2007, but their average number of employees was down by 30 per cent (to 3.0 per firm), mostly as a result of many more self-employed in the sector. But there is also a genuine increase in industries dominated by small firms, notably those providing private services to consumers. The number of employees in, for example, hotels and restaurants grew by 57 per cent between 1997 and 2007, to 109,000 spread across nearly 16,000 firms, of which over 80 per cent had between one and nine employees (SCB, 2008).

Although jobs in service firms may be fine, these firms on the average have a weaker infrastructure for health and safety than the two other main sectors, namely, manufacturing and public services (Frick, 2005). Private services have an average of 11 employees per firm versus 22 in the other sectors, with consequently weaker internal systems of health and safety management. Their jobs are more often on temporary contracts and there is a faster turnover of both personnel and firms. Managers and workers thus have less experience of the local jobs and have less time to develop cooperation, including on OHS. A lower level of unionisation in this sector results in weaker unions, which are able to appoint only one-third as many local safety representatives and one-fifth as many regional safety representatives per 1000 employees as other sectors (Frick et al., 2005). Labour inspectors visit service-sector small firms less often than those in manufacturing, and these firms rarely hire help from OHS services. Finally, the fact that revenues are predominantly in cash facilitates the use of illegal labour hire and resort to other forms of economic criminality (Holmgren, 2008).

Outsourcing reduces local OHS dialogue

With increasing fragmentation of and outsourcing by larger private-sector firms and privatisation in the public sector (also often associated with outsourcing), more and more production is done by small entrepreneurs on contract to larger and more powerful purchasers. In the procurement contracts, these purchasers often influence the conditions of production – and thus the work environment – of the entrepreneurial firms, for example in the businesses of caring, refuse collection, hauling, recycling, and rail and other public transport (SOU, 2007, ch. 4). Local managers (of the formally responsible employers) may have limited control over conditions that can result in occupational injuries and diseases. Even a good dialogue between managers and workers may then not be enough to prevent ill health. Such a dialogue is also harder to achieve. Workers in entrepreneurial firms may lose their jobs if the firms' contracts are not renewed. As with all insecure jobs, the workers undertaking them are less likely to demand improved conditions of work (Aronsson and Gustafsson, 1999).

More migrant workers, especially after EU enlargement

In general, people born outside the Nordic countries increased their share of the Swedish population by 39 per cent between 2003 and 2007 (to 1.1 million out of 9.1 million). After EU enlargement in May 2004, there was a large increase in migrant workers especially from Poland and the Baltic States, competing on the Swedish labour market (Dølvik and Eldring, 2006). Most migrants work in the small-firm-dominated sectors of private services and construction.

From the end of 2003 to the end of 2006, the registration by the tax authority of foreign firms and self-employed workers grew only slightly more than that of Swedish ones (SOU, 2007, p. 143). However, this is not a good measure of the impact of such firms and self-employed workers because EU citizens may work up three months without a work permit, and the Posting of Workers Directive (96/71/EEC) permits EU firms to operate for six months in Sweden without registration. At the same time, the number of persons with European Economic Area work permits (who officially stay and work for more than three months) tripled between 2003 and 2006 to over 10,000, mainly from Poland (Migrationsverket, 2008). To these should be added several thousand with seasonal work permits, working mainly in agriculture in south Sweden. Many also work more than three months but go back home occasionally and thus don't have to register. And immigrants from the neighbouring Nordic countries do not need work or residence permits (LO, 2004).

Some labour hire firms that specialise in migrant workers try to avoid labour taxes by claiming that their workers are living abroad and work in Sweden only on a temporary basis. This occurs notably in construction, transport, farming, forestry and fruit picking. The workers involved are placed in a more precarious situation, as they are not formally part of the Swedish labour market. However, the tax agency has revealed several of these claims to be false (LO, 2004).

A growing share of employees in small firms are thus foreigners, who have less access to information, contacts, unions, alternative jobs or other means to improve their OHS. An overview of work-related health among immigrants found a mixed situation (Schaerström, 2006). The available indicators reflect varying job situations and health outcomes between different nationalities and also between women and men immigrants. Yet the overall picture was that migrant workers in Sweden are more afflicted by injuries and disease from poor OHS than are Swedish-born workers. The immigrants often work in small enterprises from, or run by people from, their home countries. These may be perfectly legal, but the unions have found many cases in which migrant firms violate Swedish OHS and other regulations. Migrant labour also work more often in the grey zone of involuntary self-employment (as is often required and instructed by labour-hire firms that recruit them from their homelands), and as undocumented, black labour.

A similar increase in undocumented illegal labour

At the same time as migrant labour has grown, the number of asylum seekers has risen to some 27,000 per year in the 2000s. More and more of them come from outside Europe, now mainly from Iraq. The number of residence permits granted has also grown continuously during the 2000s, to 77,000 in 2007 (Migrationsverket, 2008). This is twice the average of the 1990s – except for the peak of refugees from the warring ex-Yugoslavia in 1993–4 – and three to four times the number in the 1980s.

With more refugees and other migrants, there is an increase in illegal/informal labour. These undocumented workers are now estimated to amount to some 20,000–30,000 or possibly more (Holmgren, 2008). This is only less than 1 per cent of all those working, but the rate is higher in services and construction industries. This labour is partly made up of asylum seekers waiting for decisions who have a temporary right to work but who rarely register this, in order to retain their daily allowances. There are also those already refused asylum, and those who never applied for it (or a work permit), both of whom live and work underground. If undocumented workers protest against their wages or other working conditions, most of them are likely to be deported when employers report them to the police (Holmgren, 2008).

Nearly all foreigners without papers work in industries dominated by small firms and by cash revenues, mainly in transport (taxi drivers are mostly from abroad and many lack proper papers), in hotels and restaurants, in shops, in cleaning and in construction (Holmgren, 2008). As everywhere, the conditions are worst for undocumented immigrants, who have no rights or alternatives. Because of the large share of unknown firms and workers among all the migrants (that is, not only the undocumented ones), the usual OHS statistics do not reflect their work environments, except their work fatalities. These seem to be high. For example, although they constitute only a small fraction of the workforce, some 10 per cent of fatalities befell foreign workers doing temporary work in Sweden, both in 2003 and in 2007 (AV, 2004, 2008b).

Foreign construction workers are frequently exposed to high OHS risks

OHS conditions for migrant workers in small enterprises can be further illustrated by union examples from two of the major industries employing migrant labour: construction and cleaning. After EU enlargement in May 2004, there was a rapid increase in registration of foreign self-employed construction workers, mainly from the new member states (LO, 2004). To this should be added all those who did not register. Already before that, during a six-month campaign in 2003, the construction workers unions of southernmost Sweden visited 148 sites, where they found in all 382 foreign workers. Most of these were formally self-employed, but the union considered them to be de facto employees, and therefore (at that time) working illegally in Sweden. The union also found some 40 labour-hire firms that had recruited these workers – also to large construction companies – and instructed them how to become 'self-employed'. In July 2004, 15 Portuguese construction workers were found at a mining site, recruited by a South African firm; they received half the normal wage and no benefits, and experienced poor OHS conditions The union had to threaten industrial action, first to secure a normal contract and then to make the firm pay the normal wages to the Portuguese. In August 2004 Polish workers were found removing asbestos with no protection whatsoever, which the local union stopped. In July 2003 a fatal accident to a Slovakian worker doing maintenance work in a steel mill revealed a foreign contractor that provided its workers with extremely low wages and with poor OHS and living conditions.

The number of migrant construction workers has continued to increase since then. In late 2007, the national OHS officer of the construction workers union estimated that there were around 5000–6000 foreign workers and 300–400 foreign firms operating in the greater Stockholm area alone (Johansson, 2007). Some ten firms recruited them from abroad, mainly from the new EU 10 countries. They helped the migrant workers, in their own languages, to avoid taxes and insurance by becoming 'self-employed', even when they worked at only one site doing the same job as Swedish-employed workers. To avoid employment regulation is fairly easy, as control of the foreign self-employed is both limited and difficult. However, foreign workers and self-employed frequently earn much less than they have been led to expect. In several cases they have been cheated completely of their payments, in some of which the union was able to help them. There are also many illegal, undocumented construction workers – especially in small firms, at small sites – but their number is very difficult to estimate. Small foreign firms very often operate under poor OHS conditions, for example, with few precautions against accidents or when they remove asbestos. The OHS regulations in their home countries are usually acceptable. Yet, as there is very little inspection and enforcement at home, these firms are not used to complying with regulations (see Woolfson, 2004, on the weak OHS enforcement in the EU-10).

Undocumented cleaners at the bottom of sub-contracting chains

A study by the Swedish Building Maintenance Workers Unions estimated that there are some 4000 foreign undocumented workers in the cleaning business, of which some half work in the Stockholm area (Holmgren, 2008). Around 20 yearly bankruptcies – several of which are fraudulent – usually reveal that they have exploited undocumented workers, who then frequently loose even their very low wages. Most of these firms combine a legal business with a non-legal one – where the undocumented are employed – in order to get access to some officially registered cash. Very often, these firms are at the bottom of a subcontracting chain. The primary procurers – large firms and public organizations – have pressed the price to a level, at which they may understand that the job cannot be done legally. The undocumented are lowest on the wage-ladder, with often 10–30 per cent of the minimum wage by collective agreement. Or sometimes less, or nothing when the employer refuses to pay (Holmgren, 2008).

The workload – and thus the risk of musculoskeletal injuries – is high already in the more or less legal cleaning business. Regional safety representatives regularly encounter this (Frick, 2008). For example, in a hotel the foreign-born female cleaners worked harder and harder to keep an increasing number of rooms spic and span, instead of cleaning at a lower standard or complaining to their male Swedish supervisor. A migrant woman had 80 stairs to clean, while a male migrant had only 26. She complained to the regional safety representative, who asked the manager to look at papers on the workload of all his employees. However, as no such papers existed, the supervisor realised who had squealed, and this woman was fired before the next regional safety representative visit. If she is undocumented, a foreign women can also be required to provide her boss with sexual services, in order to get enough jobs to provide for her family. If she refuses, she can be exposed to the authorities and probably deported to her homeland, where there is little else but full prostitution to live on (Frick, 2008).

More vulnerable workers in need of OHS-support

Labour market changes underline the original purpose of the regional safety representative system

The prevention of ill health at work in Sweden was never intended to be implemented mainly through state regulation and enforcement. The role of the state was limited to setting minimum standards and spreading the knowledge and implementation of them, especially through the inspection of workplaces with the highest risks. Occupational health and safety was instead meant to be managed voluntarily, by local employers. However, to be able to both demand and support a good local health and safety management, workers and their safety representatives were given rights to an extensive dialogue on the matter (Frick, 2002; Frick et al., 2005). This remains the case today, but under the current government the role of the state is to be further diminished, with reduced funding of the Work Environment Authority and for the provision of information on health and safety at work. At the same time, economic changes have weakened the abilities of small enterprises to undertake internal dialogue, even more so than when the social partners saw the need to support them through the regional safety representative system (SOU, 1972).

On the managerial side, there are more small firms. They also have less capacity to manage their work environment, as more of them are micro firms and as they expand mainly in the service sector, with a clearly weaker health and safety infrastructure. Finally, their production – and thus also their arrangements for health and safety – are to a larger extent dictated by conditions specified by large customers or suppliers. As small firm managers have less control of health and safety, their workers become less able to engage in an effective dialogue on the subject. When there are possible improvements to discuss, the workers are less able to speak up. They are more often threatened by unemployment – which now means harsher conditions – through increasing rates of temporary jobs and through the more insecure economy of their employers. Workers in small firms are also more often non-unionised, especially those with temporary jobs. At the same time, as in larger workplaces, the ever growing number of casual workers in small firms face greater risks to their health and safety than those with permanent jobs (Aronsson et al., 2002).

Although foreign workers are often not union members, Swedish unions try to uphold Swedish OHS and other labour standards. To reduce unfair labour competition through social dumping, they try to reach and support migrants and other non-members, including through informal visits by regional safety representatives and others. For example, the construction union hires interpreters (as do other unions) to reach workers from the new EU states. However, like undocumented workers, these workers frequently run when they see a regional safety representative or other union representative approaching the construction site where they work (Johansson, 2007). They are aware of their vulnerability as workers with few rights, undocumented or self-employed, and perceive union attempts to achieve collective agreements on wages at Swedish levels, OHS compliance and other labour rights as a threat because it may result in employers simply firing them or closing the firm and moving on to open another with new workers.

Future developments

There remain very many small firms with good industrial relations and high aims for their local systematic work environment management. Among these the system for regional safety representatives is still largely intact, accepted and operational. Despite its many other cuts on health and safety matters, the government has retained the level of reimbursement to the unions for their regional safety representatives. The central social partners also continue to cooperate on health and safety issues, often in industry-based committees and mainly to support small firms. There are also voluntary initiatives from larger firms to require and support health and safety standards from their subcontractors, to prevent such cases as the steel-mill fatality mentioned above. As we have seen, however, at the same time there are growing health and safety problems for small firms, and regional safety representatives confront increased difficulties in helping workers who are employed in them. It is important therefore that the latest inquiry on the work environment legislation (SOU, 2007) has proposed some changes, which may support OHS in general but particularly small firms. They include

- extension of the appointment of regional safety representatives to workplaces without a union member but where the union has – or usually has – a collective agreement with the employer (pp. 52–7);
- increasing the duty of work environment management coordination for those who commission construction work, to better implement EU Directive 92/57/EEC (on OHS in movable construction sites) (ch. 3); and
- placing a duty on those who procure services worth 1 million kronor or more, to emphasise the work environment and to make it possible for entrepreneurs to comply with work environment legislation in the tender specifications (ch. 4).

The second proposal is already being implemented, but the two others will be part of the government's discussion on the future work environment politics (Ds 2008), to which it has also invited the social partners (Barrefeldt, 2008). However, employer organisations are already opposed to the first proposal (SOU, 2007, pp. 53–4), and the third is probably also controversial. It will increase not only the legal duty but also the administrative work to procure more costly services.

At the same time, the future of the Swedish labour market model within the EU is under debate. The 2007 verdict of the European Court of Justice in the so-called Laval case (C341/05) restricts unions' rights to undertake industrial action to uphold Swedish labour standards. The case concerned the means a union may use to make foreign firms sign a Swedish collective agreement. Its outcome will have implications for how regional safety representatives are able to support health and safety standards in small firms since it influences their opportunities of access to workplaces of temporary foreign firms. A public inquiry (Direktiv, 2008) will look into the precise effects of the verdict and how the labour law may be adapted to accommodate the Swedish model of regulating working conditions more by collective agreements than by law. Already, as we have seen, workers in Sweden face increased difficulties, especially in small firms, and these have adverse implications for effective health and safety representation within them.

9
Trade Union Strategies to Support Representation on Health and Safety in Australia and the UK: Integration or Isolation?

Rebecca Loudoun and David Walters

This chapter focuses on the role of trade unions in supporting worker participation in health and safety at work. It is centred on a study of support for worker representation in Australia and the UK as seen from the perspective of union officials and other key union actors. Drawing on wider research highlighting the importance of external trade union support for effective worker representation and consultation in health and safety, it reviews the development of the trade union role in representing workers on health and safety and its regulatory institutionalisation. Within this regulatory context it examines the traditional roles performed by unions to support health and safety representatives. It acknowledges the current crisis evident in the regulatory model; and in the light of this background it discusses the views of trade union full-time officers in Australia and the UK concerning the challenges unions face in continuing to support health and safety representatives at the present time. In particular it examines how they regard the implications for trades unions of recent changes to the structure of the labour market – which make many workers harder for unions to organise on health and safety matters. It further considers their reactions to changes in the nature of the issues embraced by workplace health and safety and the extent to which they understand the subject as part of labour relations or as something separate. The chapter explores the consequences of these understandings for union health and safety strategies in the context of wider aspects of trade union organising. Responses to all these issues are compared between the UK and Australia, and the implications of the findings for trade unions are discussed.

Institutionalising the role of trades unions in workplace representation on health and safety in the UK and Australia: the development of a 'preferred model'

In the UK and Australia the introduction of measures for worker representation was one of the most important aspects of the 1970s reform of occupational health and safety regulation. As much as any other feature of the reforms of that time, it has been responsible for changing the landscape of thought and practice on managing health and safety at work. While there are differences in detail, in both countries the introduction of provisions for worker representation was an important element of the strategy to achieve 'regulated self-regulation' (Wilthagen, 1994) of occupational health and safety. In both countries provisions, primarily in the form of rights and functions for safety representatives and arrangements for joint safety committees, were significant new features of the regulatory model for health and safety. Its incorporation of trade unions and their central role in supporting safety representatives was either explicit or implicit in both cases.

Despite the recognised importance of trade unions in achieving effective worker representation on occupational health and safety (OHS), there is limited research on how these institutions seek to fulfil this role in light of recent changes in the labour market, such as the expansion of contracting, subcontracting, labour hire and other precarious forms of labour, that have had a negative impact on OHS and unions alike (Quinlan et al., 2001, pp. 335–414). Consequently the aim of this chapter is to consider the role of trade unions in supporting worker participation in health and safety using data collected in the UK and Australia from union officials and other key union actors in these countries. It begins by exploring the development of the present legislative models for trade union-supported representation on health and safety. It goes on to examine limitations of those models in light of changes to the structure and organisation of work in the last few decades, before examining how trade union officers and officials perceive and respond to these changes and where OHS fits with other priorities and aims of trade unions.

In Australia, there exists some differentiation between states and territories in their approach to these provisions; but all statutes include a duty to consult with workers (usually employees) over OHS matters, and all prescribe institutions for worker participation. These arrangements are outlined in detail by Richard Johnstone in Chapter 2 of the present collection. They range from health and safety committees only in the

Northern Territory, to committees, or representatives or other arrangements in New South Wales (NSW), and health and safety representatives and committees, with distinct but complementary functions, in the statutes of the other states. In Victoria, South Australia, the Australian Capital Territory (ACT) and the Commonwealth, health and safety representatives are vested with enforcement powers to issue provisional improvement notices (PINs) and to direct workers to cease work where there is a belief of 'an immediate risk to health and safety'. Four jurisdictions (Queensland, NSW, ACT and Victoria) now provide trade unions with rights of access to workplaces on health and safety grounds.

In the UK, the centrality of trade unions in the provisions for worker representation on health and safety was explicit from the outset. Indeed, the original statutory measures on the subject owed their existence to a vigorous campaign by the trade unions and a political deal between them and the Labour government about their inclusion in the Health and Safety at Work (HSW) Act 1974, against the recommendations of the Robens Report and fierce opposition from both the Conservative and Liberal political parties of the day (Walters, 1996). The resulting Safety Representatives and Safety Committees (SRSC) Regulations 1977, introduced under the provisions of the HSW Act in 1978, gave recognised trade unions (and no one else) rights to appoint health and safety representatives and for those representatives to request employers to establish joint health and safety committees. Although, as a result of obligations to meet European Union requirements (the EU Framework Directive 89/391), legislation was later extended by the Health and Safety (Consultation with Employees) Regulations 1996 to include measures for other employees to be consulted or represented on health and safety, these later measures merely added to the SRSC Regulations and did not replace them. They applied only in situations in which the SRSC Regulations did not already apply, and in law the latter remain, as Theo Nichols and David Walters put it in Chapter 1, 'the preferred model' for worker representation in workplaces in the UK (see also James and Walters, 1997; Walters and Nichols, 2007).

A further confirmation of the absorption of trade unions into the regulatory framework in both countries is evident from the means by which compliance with the measures on worker representation was to be achieved. In theory, these requirements are subject to the same provisions for their enforcement as other aspects of health and safety law – that is, their enforcement falls within the remit of regulatory inspectorates like the Health and Safety Executive (HSE) and the various state health and safety inspectorates in Australia. But in practice any such enforcement is rare. Following the introduction of the SRSC Regulations in the UK, the HSE issued guidance to its inspectors instructing them not to intervene in the application of the Regulations at workplaces until all the appropriate industrial relations procedures had been utilised by the employers and trade unions involved. Similarly, in most Australian states advice given to inspectors ensures they adopt at most an advisory role on the implementation of regulatory requirements for worker representation; and enforcement action on these matters is extremely rare. Legal theorists in the UK have pointed out that these and other measures of the same period were introduced on the assumption that their beneficiaries were equipped to ensure compliance themselves, without the need for significant intervention from regulatory inspectorates. Wedderburn (1980) termed this kind of provision 'auxiliary legislation' and linked it to the bargaining power of trade unions in the industrial relations situation in the UK in the 1970s. That such assumptions have proved to be inappropriate in the long term (if indeed they ever were entirely appropriate) is not the issue here. It is simply that it confirms the central position of trade unions in the institutionalisation of support for the legal framework for worker representation in health and safety in both countries from the time of their inception. In addition, in Australia the rights granted to safety representatives in some states to issue 'provisional improvement notices', which employers are legally obliged to honour and which are subject only to the authority of a regulatory inspector, provide a further step in this direction.

The operation of trade union-supported representation on health and safety

The architects of the reforms that introduced these measures on trade union-supported representation on health and safety were correct to believe they would be beneficial. There is a now a substantial body of international research indicating that objective indicators of improved OHS performance – reduced injury severity, costs or exposure to hazards – are associated with the presence of worker participation structures in firms – union presence, health and safety committees or health and safety representatives (Blewett 2001, p. 4; Frick and Walters, 1998, pp. 373–4; Johnstone, Quinlan and Walters, 2005, p. 94). Other evidence shows a positive link between trade union-supported workforce participation and compliance with OHS standards and OHS management practices (Gallagher, 2001, pp. 77–8; Gallagher, Underhill and Rimmer, 2003, pp. 71, 73; Hale and Hovden, 1998, pp. 149–51; Johnstone, Quinlan and

Walters, 2005, p. 95; Shaw and Turner, 2003, pp. vi–vii; Walters et al., 2005, p. xii). Still further studies indicate the contribution of arrangements for worker representation and consultation to raised awareness of workplace health and safety issues (Biggins and Phillips, 1991a, 1991b). Researchers have argued that trade union-supported workforce OHS participation is important to the prevention of occupational fatalities, injuries and disease because it contributes in a variety of ways to increase capacity and motivates commitment to address OHS for both workers and managers alike. Managerialist explanations for this point out that management's ability to detect and remove or minimise hazards is limited without drawing on workers' knowledge and experience (Blewett, 2001, pp. 1, 4; Frick and Walters, 1998, p. 373; Jensen, 2002; Knowles, 2006, pp. 5, 18; Walters and Frick, 2000, pp. 43–4). More pluralist or conflict-orientated explanations suggest that workers' representatives act as watchdogs on managers to ensure that the health and safety interests of their constituents are not compromised by management's focus on maximising production and profit – which may be prioritised above optimal OHS management (Walters and Frick, 2000, p. 44; Walters and Nichols, 2007). Others frame their explanations for safety representatives' actions in terms of 'knowledge activism', in which they suggest worker representatives engage in a kind of a 'political activism organised around the collection and use of a wide variety of health and safety knowledge' and avoid the marginalisation which would otherwise be brought about by professional and managerial colonisation of technical knowledge (Hall et al., 2006). At the same time it is argued that worker representatives negotiate solutions that avoid polarising dialogue between themselves and employers into industrial disputes which simply manifest conflict between labour and capital (Storey, 2005; Storey and Tucker, 2006).

Whatever theoretical construction researchers use to explain how safety representatives act effectively, there is widespread agreement that it is trade unions that provide the main source of support for their action. This is the case in relation to support from outside the workplace, especially in the form of trade union training. Extensive provision for this training is made in the UK through the TUC (Trades Union Congress) Education Service and individual trade unions' education programmes, and is also provided (to a lesser extent) in Australia through individual unions and state-level confederations.

Support comes not only in terms of delivery of training for health and safety representatives but also in the conceptualisation and design of its curriculum and content. In this respect labour education has often succeeded in transforming conventional technically or legally oriented health and safety concepts into participant-centred and experience-based materials that use labour education techniques for their delivery and adopt ways of understanding health and safety at work that are more relevant to the needs of safety representatives (Walters and Kirby, 2003). In this process, training is deliberately constructed and delivered to maximise opportunity for integrating health and safety issues into wider trade union agendas. First, the predominant pedagogy adopted in trade union health and safety courses is derived from labour education principles, and provides for the development of a particular understanding of health and safety issues grounded in workers' collective experience and values. Second, training concentrates on the role of the representative in the labour relations processes involved in workplace representation on health and safety, rather than being concerned solely with technical or legal matters. Third, training courses, and especially follow-on courses, allow opportunities for health and safety representatives to meet one another, share experiences and build up networks of relationships with one another which foster continuing support for a common understanding of collective approaches to representing the interests of workers on health and safety. Fundamental to these understandings is the idea that health and safety issues cannot be separated from those of work organisation, since successful resolution of many health and safety matters requires a deeper and wider understanding of their organisational context. It follows from this that in trade union safety representative training a strong link is forged between health and safety and other aspects of labour relations.

Trade unions also provide a main source of support for safety representatives within workplaces through the integration of their activities within wider workplace trade union organisation (Walters, 1987; Walters and Gourley, 1990; Walters and Nichols, 2007).

Current limitations to the regulatory model

This is all very well, but, as studies have also indicated, the model of workplace representation provided for by the statutory measures to improve health and safety outcomes is subject to the presence of a number of preconditions before beneficial effects are likely to occur. The changing industrial relations environment of recent years has adversely affected many of these preconditions (Walters and Nichols, 2007; James and Walters, 2005; Johnstone et al., 2005). As well as a strong legislative steer that provides rights and functions for safety representatives, these authors argue that other preconditions include demonstrable senior management commitment to both OHS and a participative approach, along with sufficient capacity to adopt and support participative OHS management and implement competent hazard/risk evaluation and control; external trade union support and consultation; and communication not only between worker representatives and managers but also between these parties and their constituencies – and not least the presence of effective autonomous worker representation at the workplace.

It follows from the above that the most complete implementation of regulatory requirements is found in larger workplaces where there is a workforce of unionised 'permanent' employees and where representatives have such facility time as is necessary for them to actively represent their members' interests. In both the UK and Australia such workplaces are now fewer in number than in the 1970s and 1980s.

Trade union membership has declined and employment has shifted from industries in which organised labour had a strong presence to those in which it is much weaker. During the same time there has also been a strong political assault on the existence and legitimacy of trade unions as the organisations representing the interests of employees. Such changes have made it more difficult for representatives to find time to be trained in or engage fully with health and safety issues and have also altered the nature and composition of their potential constituencies. Johnstone et al. (2005) point out that traditional regulatory measures have built the principal institutions for workplace participation – occupational health and safety representatives and occupational health and safety committees – around the presumption of an identifiable and relatively stable group of employees located together or in very regular contact, and working for a single employer. Many current work arrangements break with this tradition and weaken the nexus on which it is based, and therefore pose major problems for the effective use of such institutions. Problems thus created are essentially of two kinds: those that reflect incomplete or inadequate legislative coverage and those resulting from operational difficulties with existing provisions.

These features are apparent in the British case presented by Nichols and Walters in Chapter 2 and are demonstrated by reduced numbers of health and safety committees and trade union safety representatives and an increase in so-called direct methods of consultation on health and safety, linked to the declining presence and influence of trade unions in British workplaces. Although fewer such studies have been undertaken in other countries, the similarity in the broad features of change in the structure, organisation and regulation of work across most advanced market economies means that similar consequences for worker representation are likely to be felt in these countries, too. Situations in which health and safety representatives experience difficulties in 'getting things done' at the workplace also occur, and, as Nichols and Walters conclude in Chapter 2, as a result, even in workplaces in which arrangements for representation are best developed, operation often falls short of what is provided for by regulation.

Trade unions and industrial-relations researchers are of course well aware of the challenges to organised labour that are presented by the decline in trade union membership and its causes. Much effort has been spent in recent years to both understand and reverse these trends. But, remarkably, the role of health and safety issues in strategies for trade union renewal is largely unexplored in the burgeoning international literature on this subject (see for example, Fairbrother, 2002; Gall, 2003; Gospel and Wood, 2003). It was to address this gap that preliminary research reported here was conducted in Australia and the UK. It focused on trade union officers and officials, including some with responsibilities for health and safety and some with wider labour relations functions. It set out to examine: how they perceive and respond to the challenges evident in continuing to support health and safety representatives amid the change in the structure and organisation of employment and the labour market; their reactions to changes in the issues embraced by workplace health and safety; the extent to which health and safety issues are seen as part of labour relations; and, the consequences of these understandings for the organising strategies of trade union.

The scope and methods of the study

The fieldwork on which this chapter is based was undertaken during two periods, the first in Australia in 2005/2006 and the second in the UK from mid-2006 to early 2007). A horizontal study design was used in Australia, randomly selecting 14 state-based unions with members covered by the Queensland Workplace Health and Safety Act 1995 and registered with the Queensland Industrial Relations Commission. Eleven unions agreed to participate, and interviews were conducted with 20 representatives of these unions, organising workers in most key sectors of Queensland industry. Included in the study was a balance of large, medium and small unions, representing workers in a mixture of strongly and weakly unionised workplaces, with members who comprised both white-collar and blue-collar workers. Unions organising both privately owned workplaces and publicly owned workplaces were included, but the latter were over-represented in the sample. Interviews were conducted with officials in Brisbane and in two regional locations (where unions had regional offices or officials) using a semi-structured interview schedule.

Based on lessons learned in the Queensland fieldwork, and to take account of major differences of scale, a somewhat different approach was adopted in the UK. A vertical study design was used with the aim of gaining in-depth knowledge of OHS activity in a small number of unions. Interviews were conducted with 11 officials from four unions and with ten key actors in union OHS activity (the TUC, OHS trade union-approved trainers and authors of union OHS publications), using a semi-structured interview schedule similar to that used in Australia. A previously published paper based on the fieldwork undertaken in the Australia study found an apparent lack of appreciation on the part of union officials about how OHS might be used to build awareness of the role of unions among both existing and potential members (Barry and Loudoun, 2006). Part of the reason for extending the study to include fieldwork undertaken in the UK was to explore whether this interpretation of the Australian results reflected experience among trade unions in other countries.

As a means to gain a better understanding of trade union approaches to supporting health and safety representation, the interviews in both countries sought information on two broad issues: first, what union officials/officers saw as the main challenges for health and safety representatives in the modern workplace, and second, what unions did to support representatives to be able to deal effectively with such challenges. Relatedly, the position of health and safety in wider trade union strategies for their 'renewal' was explored through questions on the extent to which supporting improvement of the work environment was perceived to be central to the mission and role of trade unions and how good they were considered to be at integrating and prioritising OHS in their overall approach to representing the interests of workers.

Trade union approaches to supporting safety representatives in the modern world of work: addressing the challenges

As already acknowledged, trade union training and information constitute a prominent form of institutional support for health and safety representatives. Such provision is an important factor contributing to their effectiveness (Biggins and Holland, 1995; Culvenor et al., 2003; Walters et al., 2001). At the same time, it is evident from their published materials that in the UK and Australia both the national education and the health and safety departments of trade unions use a broad definition of health and safety in their approach to the subject and stress the importance of its integration with other aspects of labour relations (Walters and Kirby, 2003).[1] In theory, this broad-based and integrated approach should help to support representatives in dealing with the challenges of the changing world of work and at the same time help to integrate their role in trade union workplace organisation. Awareness of these issues was evident among trade union officers with health and safety responsibilities in the UK. They perceived modern workplace health and safety issues to be complex and inextricably bound up with other, labour-relations issues and felt their task to be to empower workers to solve workplace problems collectively. As one UK official put it:

> Looking for the underlying issues is difficult and there are usually IR issues involved. To solve them we need to work together and sometimes they can't be solved in one arena alone. For example work intensification, long hours and stress need a two-pronged approach.

Awareness of the interrelatedness of health and safety with other issues was evident in remarks such as:

> Health and safety usually gets caught up in other issues. An example of this is the long-hours campaign. Health and safety was a good reason to reduce the hours but it wasn't the only reason.

There was concern to prevent health and safety representatives becoming isolated:

> [We aim] to raise awareness and support for health and safety reps so that they are not working in a silo. It is about people getting actively involved in the workplace ... It is another arm, another eye.

One British trade union health and safety officer said:

> We don't generally solve health and safety problems for members. We aim to arm them with the information or tools to do it themselves through collective action. We help them to build a team. By doing

this and encouraging them to work together they know that if they stay together they can solve their problems.

It is clear from these remarks that these respondents' notion of health and safety and of health and safety representatives was far removed from the isolated 'safety technician' role sometimes attributed to or expected of health and safety representatives. Their emphasis was not on the possession of any particular technical expertise on occupational health and safety but rather on the complexity of the issues involved, their underlying links to matters of work organisation and hence the link between the task of representation on health and safety matters and other forms of trade union representation.

The challenge of addressing such complex health issues that are prominent in the modern workplace, such as those of stress and bullying, was widely acknowledged, and the tensions apparent between individual and collective responses were also identified:

> Often health and safety is seen as an individual issue and sometimes change requires a collective effort, especially when it relates to less tangible risks like stress and things that relate to the culture of the workplace. Unions need to raise this collective spirit. Health and safety reps can't do it alone.

The preferred trade union approach to addressing these challenges through collective approaches was much in evidence in British responses:

> Stress is often difficult ... because people say 'we are all under stress. Why is that person special and going off sick?' The challenge is to unpack the individual stress claim and get to the underlying, collective issues.

These wider conceptualisations of occupational health and safety were therefore helping to integrate the subject into wider labour relations concerns and, as one regional official put it, unions needed to *recognise and embrace* the things that make the union relevant to a wider cross section of the labour market:

> Many officials are still stuck in the 1970s and think that union work is all about pay. We need to raise the priority of health and safety,

particularly work–life balance, quality of life etc. These are the things that new workers care about.

In a similar vein a trade union trainer commented:

> Health and safety can bring women into the union when they wouldn't have come for other reasons (like pay).

At the same time these wider conceptualisations were also seen as presenting challenges. For example, stress-related illnesses were perceived to be influenced by the different coping levels of individuals. What one person could deal with another could not. Often there wasn't a clear right or wrong, as there might be with more traditional health and safety issues, and such matters were sometimes difficult to treat as collective issues:

> [Health and safety] is difficult when the issue is open to interpretation. Unless there is a clear right or wrong people don't like to complain. They try to settle it themselves without contacting the union or going through formal channels.

It was not so obvious that officials working in capacities other than education or health and safety accepted the same degree of integration. Although it was recognised that issues such as psychosocial stress and harassment were now firmly on the health and safety agenda, and their consideration required negotiation over the organisation, timing and pace of work, which placed pressure on the traditional separation of OHS from industrial relations, most officials still didn't see OHS support as their role and continued to seek others to whom they could pass these issues on. They were happy to work with health and safety representatives but not to incorporate OHS into their own role when negotiating on the organisation of work.

The Australian findings were similar to this, although they differed to the extent that unions were still largely reactive on health and safety matters – they dealt with health and safety issues as they occurred rather than proactively anticipating them as appeared to be the case among some UK trade union officers. Many Australian union officials were of the view that it was up to members to bring OHS issues to them as a collective grievance. It was only then that they would get involved:

> There can be no doubt that health and safety is important to us and to our members but we are not onsite so we don't know what the

day-to-day problems are. Of course there are the obvious ones like working at heights but we need members to keep us informed of any new things they are worried about. We can't be everywhere at once and we can't fix everything at once.

It was difficult to get involved because according to many officials workers, particularly those in regional areas, had a low level of awareness about OSH issues. They were often hamstrung by what they saw as the complexity of OHS issues, which was also perhaps a reflection of and a contributor to the lack of resources they invested in OHS training, education and promotion. In Australia, then, while there was evidence of a similar approach to the broadening and integration of health and safety concepts among trade union officials to that found in the UK, the strategy of integration was often regarded more as a means to manage scarce resources (that is, to relieve officials from dealing with OHS issues) than as a strategy to educate and promote better OHS outcomes. As one official noted:

> OHS doesn't take a bottom ranking but I do try to get reps to acquire the knowledge to deal with it themselves. The OHS committee at the workplace is often window dressing and I just don't have the time.

The emergence of 'new' health and safety issues was also seen as a resource issue in Australia, as quite often there was no one onto whom such issues could be passed:

> Hazards such as broken air conditioning and fire alarms are more easily identified as issues, compared to other hazards like violence and bullying. The union needs more resources on how to organise against psychiatric damage like violence, overwork, bullying and work intensification.

The question of trade union resourcing for health and safety was also a concern for some UK officials. Indeed, most officials interviewed in Queensland and the UK identified a need for more union resources on health and safety to give them a greater capacity to identify issues and therefore help health and safety representatives – this was felt to be especially necessary to enable them to address less overt hazards. As one UK official put it:

> Looking at the resources that unions put into OHS they don't have paid officials in the regions (nor does the TUC) that are dedicated to

OHS. Without this support and skills people skirt around issues they don't feel comfortable with or they don't have somewhere to turn.

There were therefore some differences of approach to health and safety and to health and safety representatives between trade union officials with responsibilities for health and safety and officials with other responsibilities, as well as between British and Australian trade union officials. Broadly speaking, however, there was widespread recognition of the changed nature of work and its health and safety consequences, as well as the link between health and safety and other trade union issues. There was acknowledgement of the need to integrate the role of health and safety representatives into other aspects of trade union organisation at the workplace and of the potential interest of workers in matters affecting their health or safety.

The major challenges for effective representation on health and safety in modern workplaces in both the UK and Australia concern how to address the decline of the preconditions for its effectiveness. In particular, the reduced trade union presence in many workplaces, the increased fragmentation of the structure and organisation of work and the changed nature of the labour force present major hurdles for effective representation on health and safety. The way trade union officials conceptualise the subject and their approaches to health and safety representatives outlined in this section is relevant to understanding how they address these challenges. A more explicit measure, however, might be found in the approaches of trade union officials towards the role of health and safety in their organising strategies. These are explored in the following section.

Health and safety and trade union organising

Trade union organising strategies embrace a wide range of activities with different focuses under the general remit of achieving renewal through the deployment of techniques specifically developed to improve recruitment of trade unions members, recognition and workplace/branch organisation. In the past decade there has been considerable interest among trade unions and labour-relations researchers concerning the effectiveness of these methods in reversing the decline in membership and influence experienced by trade unions in many modern neo-liberal market economies. Of central interest for the future of health and safety representation, therefore, is the extent to which trade union officials regard health and safety as having a role in their organising strategies and how health and safety representatives are supported in their activities by such strategies. We conducted our interviews in the UK with the aim of exploring these issues somewhat more explicitly than had been the case during the previous Australian fieldwork.

In 2008, the TUC published its guide to organising around health and safety. It set out to show how union organisers, officers and health and safety representatives could use health and safety as a tool in campaigns for union recognition as well as to be better organised in already unionised workplaces. In addition to new guidance and training aimed at union organisers, it contained revised editions of previous TUC publications on organising that were aimed at health and safety representatives. Although the fieldwork for the present study was completed sometime before this publication, the line it takes was already evident in the views expressed by members of the TUC's organising and recruitment team who were responsible for its production (TUC, 2008):

> Health and safety has not been an explicit element of our push for organising. In 1998 we launched new unionism and the academy. OHS is only coming on board now, which is in part because we are progressing in how we approach organising and becoming more sophisticated – branching out in different directions. When the organising push started it was a small area that operated alone while the other departments got on with what they had always been doing. Now organising forms part of what everyone does.

> Organising is no longer a bolt on. It is built into everything we do. ... Our approach is not perfect but it is progressing. Health and safety and organising complement each other. Better organisation makes for better OHS and vice versa.

Despite this very clear line from the TUC, our interviews with regional officials responsible for organising indicated that many were unaware of the previous versions of the TUC organising materials for health and safety, and there were considerably mixed views concerning its place in the organising agenda and some way to go before its role was widely accepted in practice:

> When looking at health and safety, it is not high on members' priority list unless an accident has happened recently. This makes it very difficult to organise around. If you get people at the workplace together and ask them what is going on/worrying them at their workplace they will mention pay and working conditions before health and safety ... if they mention health and safety at all.

On the role of health and safety in recruiting and organising among 'hard to reach' groups another said:

> With migrant workers we have tried to use health and safety to organise. They usually have poor working conditions with no induction/health and safety information in their language etc. Unions try to use the health and safety message in these workplaces and with these workers as they see it as a button to push that might be effective but things like English language skills seem to work better. They worry what their employer will think. And they tend to move around a lot too.

Other officials appeared to appreciate the potential of health and safety as an organising/awareness raising issue, but bemoaned the extent to which it was utilised as such in approaches to organising:

> Health and safety is a good hook to organise and recruit around in the workplace but I don't think we do a good enough job in using it especially when it comes to campaigns. We usually don't emphasise the health and safety angle. This is a shame because the imagination can be sparked off quite easily in health and safety. When accidents happen at the workplace we don't advertise/publicise the issue and use it to recruit well enough. Again this is a shame. When we go to schools to talk about unions when we mention health and safety the lights go on. We are simply not talking about it enough. The potential is there but we need to raise its profile.

Some union officials suggested that the complexity of the issues involved in health and safety made it difficult to use as an organising issue. One linked this to the limited extent of training on these issues for organisers:

> Our organisers only receive minimal training education on health and safety (maybe a couple of hours) and vice versa with safety reps on organising so we have some problems with workplace reps thinking that health and safety is all a bit technical for them – something they need a specialist for.

On health and safety and organising recruitment another suggested:

> Health and safety is actually easier to use for recruitment when there are clear guidelines. It is more difficult when the issue is open to interpretation. Unless there is a clear right or wrong people don't like

to complain. They try to settle it themselves without contacting the union or going through formal channels. This is difficult for recruitment because quite often they are not aware of their rights.

and

> Very few people bring health and safety to our attention and they tend not to stick together. The focus is more on wages and holidays, particularly with the youngsters.

Other union officials suggested that health and safety representatives were themselves a barrier to organising because they did not regard this function as part of their role, as the following several quotations make clear:

> Some health and safety reps don't see themselves as active trade unionists and they want to keep it that way – it can be hard to turn this around. They don't see recruitment as part of their job. They see health and safety as separate from mainstream union activity.

> Reps don't necessarily see themselves as forming a part of the organising agenda. They tend to see this as separate unless they are the rep and the stewards.

> They are usually a bit bemused when we get to that point but it usually turns out to be a positive experience.

> The organising unit is often seen by older HSRs [health and safety representatives] as something they don't have to do. We have to change this perception.

> Safety reps generally don't want to organise and we don't have any dedicated organisers in Wales, which makes it hard for them and us.

But others saw the problem as being to do with lack of attention paid to the link between health and safety and organising by the union itself:

> Support from individual unions to raise the profile of health and safety would help a lot [to recruit using health and safety] – in general unions have failed to own health and safety.

and

> For my work (organising campaigns) it is a problem that we don't have more people with expertise in health and safety. There is no real structure.

Despite the somewhat negative interpretations of the relationship between health and safety and organising suggested by the above quotations, other trade union regional officials were considerably more positive. Experience had clearly suggested to some that health and safety was a useful organising tool. For example:

> Health and safety is one of the key things we deliver to new members. It fits the organising role very well.

Reasons that officials gave for feeling positive about using health and safety as an organising issue concerned its perceived legitimacy as an issue for trade union action:

> Health and safety is a good area because most people see it as a legitimate issue for unions to operate in and as such people are more amenable to joining a union for health and safety than for other reasons particularly if they are generally opposed to unions for whatever reason.

The support of the law was also seen as advantageous here:

> Health and safety is good for recruiting because when the times are tough this door is usually still open to access the workplace. The law is very much on our side when it comes to accessing and fixing health and safety problems.

Its appeal to younger workers was frequently noted:

> Health and safety is an easy win for us and the shop stewards so it is great for organising. Younger people and people new to the industry tend to respond to health and safety.

Some officials also shared a positive interpretation of its technical aspects in relation to recruitment:

> The technical aspects of health and safety are very good for recruiting. They are usually winnable and have a clear right/wrong answer. The people end of it is much harder. This is not to say that it is not as good or effective but it is harder.

A somewhat mixed picture of the perceptions and attitudes of trade union regional and local officials regarding the role of health and safety in organising emerges from these findings in the UK. It seems evident that, while a positive relationship between health and safety and organising is clearly set out in national-level policy and training materials from the TUC, practice at regional and local levels is considerably more complex. Many officials do not see obvious advantages to the incorporation of health and safety issues into their organising agendas, and some do not appear to regard health and safety representatives as helpful in relation to organising. Nor do they perceive a broader-based understanding of health and safety issues as particularly useful, often regarding it as resulting in concepts that are too complex to aid building support for collective actions. At the same time, others recognise the potential of health and safety issues to appeal to a new membership among young workers, women and other groups which trade unions are anxious to recruit. While officials often recognise the complex nature of health and safety issues, they believe such complexity makes these issues more rather than less useful ones to organise around, because they are tangible consequences of managerial strategies about which workers are concerned. There is also a view that some health and safety issues are winnable ones because there are legal or other standards to which they can be linked; moreover, they are issues on which it is possible for trade unions to legitimately occupy the 'moral high ground'.

These paradoxes suggest that, although the thinking about health and safety and organising among trade union regional and local officials in the UK is more developed and proactive than that previously found among their counterparts in Australia, it is still far from the case that health and safety issues feature centrally in the organising practices followed by the officials in the four trade unions studied in the UK. The accounts of many of these organisers also suggest a major conceptual gulf between their attitudes towards health and safety representatives and the literature that portrays these representatives as workplace activists who use work environment issues as the material around which they forge a form of political identity and action (see for example Hall et al., 2006).

Health and safety, organising and the future of trade union representation on health and safety

In our study we found some differences in the approaches of the Australian and UK union officials who participated. Broadly speaking, in the UK we found health and safety to be somewhat more integrated into the thinking of officials on labour relations issues and on organising more generally. British officials were also more proactive concerning its potential role in these matters than their Australian counterparts. To some extent this may have been a function of more developed union policies and training materials on these issues nationally in the UK than in Queensland/Australia. But it is also possibly a reflection of differences in the development and current regulatory position on representation on health and safety between the UK and Queensland. As Johnstone makes clear in Chapter 2, the development and present form of regulation in different Australian states place different emphases on what constitutes the regulatory model at state level. In Queensland, for some time the safety committee rather than safety representatives were its primary focus. Further, differences may also reflect practical difficulties involved in resourcing collective labour organisation in the large-area/low population-density conditions encountered in Queensland.

But the most significant feature of our findings concerns their very mixed nature. This was especially true of the British experience. A positive interpretation of this, such as given by some participants in the study, is that the relationship between the representative function on health and safety and wider strategies for organising trade union renewal is a developing one. It is one understood by many trade union health and safety activists; and there are signs that some trade union organisers have also recognised its potential. Supported by the provision of trade union training on health and safety and organising, such as that introduced by the TUC Education Service, which is aimed at both organisers and health and safety representatives, the recognition of this potential is likely to grow, and in time health and safety may well take a more central position in the organising agenda than is currently the case in practice. This in turn should help to support and sustain the role of health and safety representatives in the changing world of work.

Set against this optimistic view is another possible interpretation of the findings in which health and safety issues are not seen as central to the remit of organisers, among whom there is little appetite for complex conceptualisations concerning the usefulness of the relationship between workers' health experiences and underlying work organisational issues. Health and safety representatives are themselves regarded as somewhat of a hindrance to organising – because they frequently view such matters as not a part of their role. The relevance of health and safety issues to organisers with this view is at best peripheral, reactive and opportunistic, with little sign of its more systematic inclusion on an organising agenda of the future. In the best-case scenarios under this interpretation, organisers use health and safety issues only when they are simple, straightforward and winnable, or more inadvertently as unacknowledged aspects of other issues with which they are more comfortable.

Training, and communication between health and safety activists, trade union officers with health and safety responsibilities and trade union organisers, might address some of these gaps over time, improving organisers' understanding of the relevance and usefulness of health and safety on the one hand and the usefulness of organising for health and safety representatives on the other. However, given the widely felt perceptions concerning the limited resources of trade unions to invest in such support, as well as the possibility of a 'training overload' already experienced by organisers, it is perhaps unrealistic to expect it to achieve significant change in the short term.

These apparent contradictions in our findings should not be particularly surprising. At one level they are simply a refection of the heterogeneous nature of the subject matter of occupational health and safety and of the many complex approaches to addressing it. On another level, the findings are also quite a close approximation of those reported in other recent studies of the more general impact of organising strategies on trades unions. These studies identify inherent tensions in the concept of 'organising' that influence its priorities and effects on trade union renewal overall (Holgate and Simms, 2008). More specifically, researchers have pointed to the continuing dominance of interest in traditional trade union functions such as pay and conditions in organising approaches, with less emphasis on the significance of new issues of which health and safety may be an aspect, such as work organisation, membership diversity and human resource-management issues. They further identify a conservatism within trade unions that has limited the impact of organising in terms of its scale and the sectors in which it has been felt (Heery and Simms, 2008). These factors also influence different strategies on organising that are pursued by different trade unions. Some unions, for example follow, the TUC model in which organising plays a role across a range of issues and practices at workplace and branch levels, while others use it in a more restricted capacity in relation to recruitment of members. Organising may also be seen as an issue on which national-level policy decisions are required, while health and safety enjoys more room for local decisions. As such, there are sometimes difficulties experienced when attempting to mix the two. Although studies identifying these issues make little mention of health and safety, their findings imply that, even if it were to be included more systematically within trade union organising strategies than is currently the case, its impact would by definition be limited by the same barriers that have been shown to constrain organising itself.

Our findings demonstrate an emerging awareness among some trade union full-time officials concerning underlying connections between health and safety issues and other more 'mainstream' labour relations matters, as well as the potential for a more systematic focus on these same issues in wider trade union approaches to recruitment, recognition and the support of workplace representation. Of course, such awareness and inclusion will not fully address the challenges to sustaining the trade union-supported model of worker representation on health and safety issues that has been the basis for effectiveness since its introduction in the 1970s. These challenges have been created by a shift in the balance of power between labour and capital in the ensuing decades, leading to the breakdown of the post-war compromise and a parallel withdrawal of the state from its regulation of employment matters in most advanced market economies.

The resulting political, economic and regulatory landscape in which representation on health and safety currently operates is entirely changed from that which was in place at the time the existing measures were introduced. Not least, in policy terms an individualisation of responsibility and marginalisation of collectivism has occurred across a range of issues including those concerning the health effects of work. Current government and employer thinking on these issues in the UK, for example, presents a remarkably unified conceptualisation of the health benefits of work and a highly individualised approach to keeping workers in work through strategies which make it more difficult for them to take time off on the grounds of ill health and which promote individualised approaches to addressing the physical, mental and emotional consequences of coping with organisational demands imposed by employers. The rhetoric accompanying this thinking focuses on influencing the actions of the victims of these demands while largely ignoring their well-established organisational causes. In such a scenario it seems important that there should be greater awareness of the health consequences of the increased freedom of employers to impose demands on labour, coupled with an appreciation of the nature of these health effects (stress and other 'non-traditional' occupational health effects) and their causes, aligned with a stronger link to wider trade union organising approaches. Signs of such awareness are demonstrated in the views of many of the trade union officials in the present study.

On its own, this will not solve the crisis facing representation on health and safety, nor will it necessarily deal specifically with such matters as increasing the rate of recruitment of trade union members and trade union recognition, or significantly improve access to representation for the large proportion of the workforce employed in small firms, or lead to better representation of the interests of the employees of contractors and subcontractors or agency workers on multi-employer worksites. Gall (2006), in his analysis of the limitations to trade union organising strategies in the UK, points to the continued dependency of trade union and collective bargaining on employer and state support. But he also suggests that such dependency (and the obverse strategies of independence also pursued by unions) is not static and argues that its dynamic and changing nature and its subjection to external political and economic conditions may have positive connotations, as well as the negative ones that are more obvious in the current climate. He argues (2006, p. 235):

> This suggests that trade union power can again be created and without specifying the particular form, lay activists are the central mechanism by which this can be achieved.

By definition, health and safety representatives are 'lay activists'. However, it is apparent that within trade unions and among theorists of trade union reform they may still be largely unrecognised as such.

More pragmatically, the challenges necessitate consideration of a range of complementary strategies at national and local levels that help representation relate to emerging work scenarios. They also imply the need to explore new alliances at sectoral, local and other levels to promote representation and consultation in workplaces, and especially how such alliances might be used to help extend representation and consultation to hard-to-reach groups of workers. For example, recent American literature on trade unions and organising among migrant workers points to innovative ways in which traditional barriers to unionisation might be addressed (Milkman and Voss, 2004). Other American writings on the role of alliances between organised labour and communities engaged in environmental activism contain further important messages that are relevant to sustaining representation on health and safety (Obach, 2004; Estabrook, 2007). In Europe, the strategies of trade unions in southern European countries that attempt to link union actions to wider social mobilisation are relevant because issues of health and well-being are often fundamental aspirations of such actions (Kelly and Frege, 2004).

Nevertheless, as Lax (2006) argues:

> The source of workplace injuries and illnesses can be found in the social structure, and more specifically in a capitalist system that requires all employers to continually strive to maximize profits in order to survive, at the expense of other considerations. Workers resist, struggling for alternatives that meet their needs. The success of their struggle depends on power, class power ... social change, including the dramatic diminishment of working class power, has transformed the social landscape offering new constraints and opportunities for action. The fundamental nature and dynamic of capitalism has not changed, however, and it remains necessary to understand that nature and dynamic to understand how and why workers continue to get injured and ill on the job. With that understanding, change becomes possible.

This brings us back to the central focus of the fieldwork on which this chapter has been based. It illustrates why raising the profile of the health consequences of modern work organisation and its implications for the quality of workers' lives – and in so doing, demonstrating that organised labour and workplace health and safety representatives can play a role in limiting the freedom of employers to harm people in these ways – is a necessary step for organised labour in the direction of its renewal.

Notes

1. For Australia see, for example, the website of the Australian Council of Trade Unions http://www.actu.asn.au/HelpatWork/OccupationalHealthSafety/Helpandresources/OHSFactSheets.aspx and that of Victoria Trades Hall Council http://www.ohsrep.org.au/hazards/index.cfm.

10
Worker Representation and Health and Safety: Reflections on the Past, Present and Future

Phil James

It is a truism to say that work has potential implications, both positive and negative, for the physical and mental well-being of those who undertake it. It is similarly clear that there is nothing inevitable about how far these implications are realised and hence the extent to which work serves to harm or enrich the lives of those undertaking paid labour. Rather, much depends on both the priority accorded to the protection and advancement of worker health and safety on the one hand, and the degree to which effective systems of management are put in place to support the achievement of these objectives on the other.

As a number of chapters in this volume, notably Chapter 1 by Theo Nichols and David Walters, have demonstrated, worker representation on workplace health and safety matters can make an important contribution in both of these areas – both by prompting employers to increase the importance attached to it and by improving the manner in which they manage it, at least where this representation is provided through channels independent of management such as trade unions. Indeed, taken together these chapters serve to paint a picture in which such representation constitutes a fundamental mechanism of protection for the health and safety of workers.

Against this background, the foregoing contributions have focused attention on a range of issues relating to worker representation in the area of workplace health and safety with a view, in particular, to shedding light on the development and nature of the legislative frameworks put in place internationally to support it, the extent of their implementation and their continued appropriateness to the world of work which exists today. In doing so, they have also, both directly and indirectly, by comparative implication served to highlight weaknesses in the content and the operation of current regulatory provisions in the area, possible means of addressing these weaknesses, and the political and economic barriers that would need to be confronted and overcome for such reforms to be achieved. These reforms, by logical extension from the evidence on the role that worker representation can play in enhancing standards of workplace health and safety, could potentially do much to reduce the scale of work-related harm suffered by workers, as well the associated financial costs borne by them, their families and society more widely.

In this concluding chapter this reform agenda constitutes the central theme of interest. However, before addressing it directly, the chapter initially reviews the key points to have emerged from the preceding chapters with regard to the rights of representation over health and safety matters that workers currently have, how far they in practice have access to, and are able to utilise, these rights, and the factors which influence this utilisation. The intention being to thereby provide an evidence-based framework within which the subsequent discussion of possible reforms can be located.

Rights of representation

The rights of health and safety representation that exist within the countries examined in the book's contributions are, for the most part, the product of legislative interventions that occurred over the past half a century. In most cases, such as Canada, Australia, Sweden and Britain, the rights concerned stemmed from regulatory initiatives dating back primarily to the 1970s. These initiatives have subsequently undergone one or more waves of reform prompted, variously, by a number of factors, notably shifts in the political power of national union movements and related changes in government regulatory policies and philosophies, and labour market changes, as well as European Community legislation. In contrast, in some Baltic States and countries such as Spain the regulatory provisions are of much more recent origin and can primarily be traced back to the need of these countries to bring their domestic laws into the line with the requirements of the European Union health and safety Framework Directive relating to worker consultation and participation.

These various influences can also be seen to have prompted processes of regulatory convergence and divergence. Differing social and economic dynamics, for example, have led to the legal specification of different forms of representation in countries and in different parts of Australia and Canada, and to variations in the range of employers to whom the relevant legal requirements apply and, albeit to a lesser extent, the degree to which these arrangements extend to cover 'non-standard', or contingent, types of labour. At the same time, as the co-editors note in their introductory chapter, there is a good deal of similarity as to the rights provided to representatives; although here too some marked differences exist.

Representative forms

Marked differences have been revealed in legislative provision on worker representation on workplace health and safety in different jurisdictions. In Spain, for example, exclusive reliance is placed on the role of safety representatives, while in France the same is true for health and safety committees. Most provinces in Canada, meanwhile, provide for both these channels of representation to be utilised, and this is also true of most Australian territories and states and the British Safety Representatives and Safety Committees (SRSC) Regulations 1977; although sometimes these different channels are used to provide alternative mechanisms of representation and in other cases provision is made, as with SRSC regulations, for them to be utilised alongside each other.

These differences in representative forms exist alongside further ones concerning the role played by trade unions in the appointment of worker representatives, be these safety representatives or health and safety committee members. Thus, in some jurisdictions unions are accorded a formal role in the election or appointment of representatives, while in others they are not. In France, for example, where unions are present, only union representatives can stand as candidates during the first ballots held to elect health and safety committee members, in Britain recognised unions under the 1977 regulations are empowered to appoint safety representatives, and in the Australian state of New South Wales it is possible for unions and employers to develop representative arrangements that are distinct from those otherwise mandated. The predominant pattern, however, is for unions, in the words of Richard Johnstone in Chapter 2, to not, or no longer, possess such a 'privileged role' and for reliance to therefore be placed on processes of general workforce election.

This predominant pattern of formal 'union exclusion' from health and safety representation, in turn, exists alongside a general lack of any formally mandated linkages to broader labour relations structures – although these may, of course, in practice exist in particular workplaces on a non-mandated basis. Thus, of the jurisdictions examined in this volume, only France makes provision for such a linkage through providing for health and safety committees to be a 'sub-section' of works councils.

Coverage of representative arrangements

In none of the legal jurisdictions reviewed does the coverage of the statutory provisions relating to health and safety representation extend to cover all employers and therefore workplaces. In Britain, this situation exists because, as Nichols and Walters note in their chapter, in the absence of recognised unions existing and choosing to appoint safety representatives, employers are provided with discretion as to whether they consult workers 'directly' or via elected representatives. In other jurisdictions limited coverage mainly exists as a result of the use of thresholds, which act to restrict the application of the relevant regulatory provisions to workplaces above a particular size. Thresholds which, as Laurent Vogel and David Walters note in Chapter 5, vary considerably, ranging from five employees in the case of Sweden to 50 in France, Belgium and Bulgaria.

The reference in these size thresholds to 'employees' also serves to highlight another common feature of most legislative frameworks for representation. This is that, definitionally, they do not extend to encompass rights of representation to those who, in Anglo-Saxon legal terms, do not work under a contract of employment but rather are engaged under other types of employment relationship, such as self-employment and agency working. This means that, where such size thresholds exist, the presence of such workers can, in the case of small workplaces, potentially work against the establishment of health and safety representation by enabling employers to keep the directly employed workforce below the specified number. In a similar vein, generally the frameworks in place do not provide in situations of on-site subcontracting for representative arrangements to be established which cover both those employed by the host employer and those employed by the contractors that are present. However, as Richard Johnstone again highlights in Chapter 2, some limited action has been taken in several Australian states to at least facilitate employers and unions taking action to establish arrangements of this type.

A further point to note is that in a number of countries employers are not obliged to establish representative arrangements of the prescribed type unless 'requested to do so'. This, for example, is the case in all Australian territories and states other than Queensland.

In some parts of the Australian system, as well as Sweden, statutory provisions do, however, exist under which trade union representatives not employed at workplaces can gain access to them to take up health and safety issues with the employer. In New South Wales, for example, an appropriately authorised representative can gain entry to a workplace 'for the purpose of investigating any suspected contravention of the occupational health and safety legislation', and similar provisions also exist in Victoria, Queensland and the ACT. Meanwhile in Sweden, as Kaj Frick details in Chapter 8, unions are legally entitled to appoint regional safety representatives in all sectors of the economy who are empowered to carry out inspections, and other activities, in small firms where there is at least one union member and no joint health and safety committee exists.

Powers of representatives

As already noted, the powers of action accorded to worker representatives are, for the most part, very similar and typically encompass the right to be consulted by employers and further rights relating to the receipt of training, the provision of information from employers, engagement with government inspectors, and the carrying out of inspections. In addition, representatives invariably have entitlements to paid time off to undergo training and carry out their activities, and are afforded protection against being victimised by employers as a result of the work they undertake. They also, sometimes, have a right to draw on the expertise of outside consultants and experts and to order that dangerous work cease.

In contrast, representatives invariably do not have the power to insist that employers engage in negotiation, and hence a degree of co-decision making, over health and safety issues. As a result, the ultimate power to decide whether workplace health and safety arrangements need to be improved typically resides with employers, subject to any alternative view expressed by a government inspector.

Exceptions to this general picture of employer decision-making authority do, however, exist in Australia. Thus, here, most legal jurisdictions vest health and safety representatives with the power to issue 'provisional improvement notices' where it is believed that employer arrangements do not meet legal requirements. The employer is then obliged to take the remedial action required by such a notice unless it is set aside by an inspector, or face subsequent enforcement action.

Implementation of the legislative systems

The chapters in this volume have, then, highlighted the fact that universal rights of workers to health and safety representation are notable by their absence. They have also served to highlight another point, namely, that the implementation of the rights that do exist is invariably far from complete.

This incomplete implementation is shown to take two forms: first, the non-establishment of representative arrangements of the legally provided types, as a result of either employer non-compliance or a failure on the part of workers/unions to 'trigger' their utilisation; second, a failure of arrangements to operate in a way that provides workers with access to the full range of representation rights which are legally available.

From the evidence provided these problems of non-implementation would seem to be a product of a number of factors. First, as Wayne Lewchuck and his colleagues highlight in Chapter 6, the rise in employer power which has generally occurred over the last three decades has affected the ability and confidence of workers and unions to take full, or even partial, advantage of the legislative rights to representation that have been put in place. Second, and relatedly, there seems to be a (growing) lack of commitment of employers both to the principle of worker representation and to compliance with the legal frameworks that exist. A lack of interest on the part of workers and unions in operationalising their rights more effectively would, unfortunately, appear to be a third factor.

In practice, of course, these three sets of influences are interrelated. Workers, for example, as the analysis of health and safety in the Baltic states of Latvia, Estonia and Lithuania by Charles Woolfson and colleagues suggests, are unlikely to exhibit strong support for worker representation in a context where they feel it would be unlikely to yield beneficial outcomes and, indeed, may expose vulnerable individuals to employer hostility. Similarly, the level of employer commitment to worker representation clearly cannot sensibly be explained without reference to the degree to which employers face pressure from workers and unions to engage with it.

It also needs to be noted that these influences, as the earlier analyses again highlight, have been compounded by another, namely, the limited degree to which government inspectorates have sought to ensure employer compliance with the prescribed statutory provisions. Such a lack of inspectorate action is, of course, a more general problem that surrounds the implementation of health and safety laws across the industrialised world as a result of employer-friendly enforcement approaches and the limited resources that inspectorates have both to visit work-places and to pursue enforcement actions. In the case of worker representation, however, these general problems have been compounded by a more specific one that is also highlighted in Chapter 6 by Wayne Lewchuck and colleagues. This is the way that the widespread adoption of a regulatory philosophy of 'regulated self-regulation', under which employers in collaboration with their workforces are identified as the primary mechanism for maintaining and improving standards of work-place health and safety, has acted to downplay the role to be played by external accountability through inspections and associated enforce-ment action. This consequence can, perhaps, be viewed as an outcome of two forces: first, a view of the type expressed in the 1972 report of the British Robens Committee to the effect that there are major limi-tations on the ability of such external regulation to secure improve-ments in the way in which health and safety is managed (Robens, 1972); and second, the value of this philosophy in terms of legitimising neo-liberal-based policies aimed at reducing the regulatory burden on employers, even though the evidence to support it is notable by its absence (Woolf, 1973), and international evidence indicates that inspec-tion and associated enforcement action works in terms of improving standards of worker protection (Davis, 2004).

In short, the widespread partial implementation or non-implementation of the legal frameworks for worker representation in health and safety can be viewed as a product of a lack of the preconditions that existing evidence indicates act to support their effective operationalisation (Walters and Nichols, 2007). Preconditions that encompass strong employer commitment to such representation; worker access to inde-pendent, and ideally trade union-based, representation; adequate con-nections between representatives and (a) those they represent and (b) wider structures of labour representation; and the presence of meaning-ful systems of external monitoring and enforcement.

Enhancing health and safety representation

From the foregoing discussion three points emerge clearly. The first of these is that worker representation constitutes a potentially important means of protecting workers from injury and other forms of harm flowing from their work activities – a virtue that cannot be overstated given the sheer scale of the harm experienced by workers internation-ally and the costs, both financial and emotional, that this harm imposes not only the victims of it but their families. The second is that across the industrialised world, with the notable exception of the United States, legal frameworks have been put in place to support such representation. The third is that there is clear scope to improve both the content of these frameworks and the extent to which they are in practice implemented.

This section takes this last observation as its starting point and considers what can be done to improve both the legal rights of access to health and safety representation that workers have and the extent to which these rights are able to be utilised and thereby adequately support the central objective that they are intended to serve, namely the protection of the physical and mental well-being of workers. It does so through an exploration of three, inevitably related, themes: rights to representation; mechanisms of enforcement; and linkages to wider employment relations structures.

Rights to representation

At the heart of the legal frameworks which exist to support worker health and safety representation, as the chapters in this volume amply demonstrate, resides a (moral) paradox. On the one hand, such frameworks exist because, presumably, representation of this type is seen as desirable and therefore to constitute 'a good'. On the other hand, none provides workers with universal access to such representation. Instead, they limit this access by excluding smaller workplaces and certain categories of workers from their coverage, and often place the onus on workers or their unions, rather than employers, to initiate legally specified forms of representation.

These restrictions related to different categories of worker and size of organisation are logically indefensible in the context of a wider political acceptance of the value of worker representation, at least in terms of the fact that laws have been promulgated and maintained. They are equally indefensible when it is borne in mind that (a) injury rates in smaller workplaces and organisations have generally been found to be higher than in larger workplaces (Nichols et al., 1995; Walters, 2001), (b) there is clear evidence that levels of work-related harm are worse among various categories of contingent labour and in situations of subcontracting (Quinlan et al., 2001; James et al., 2007), and (c) rates of injury are worse, as Theo Nichols and David Walters demonstrate in Chapter 1, where health and safety arrangements are determined by management alone.

Action is consequently needed to address both of these sources of weakness. In the case of the obligation of employers to initiate the establishment of legally specified mechanisms of representation, this clearly could be easily legislated for. Care would, however, need to be taken to ensure that workers and their unions had sufficient influence over the precise arrangements put in place by employers in pursuit of compliance with the legally mandated mechanisms of representation – an outcome that could be achieved by the allocation to them of rights of co-determination and hence veto.

The restriction of representation rights to workplaces above a given size and those that work under 'contracts of employment' could similarly be overcome by relatively straightforward legal reforms. For example, this could be accomplished through the abolition of the former and specifying that representative arrangements, including the powers possessed by worker representatives, were extended to cover all those who are engaged in undertaking work on an employer's premises, no matter on what legal basis this is done. Indeed, this last change would have the benefit of bringing within the scope of representative arrangements not only those working on a casual or self-employed basis, but those who are supplied by temporary agencies and work for on-site subcontractors. This change would, though, need to be buttressed by requirements obliging such organisations to cooperate with worker representatives based in the host organisation and to comply with any rules and procedures promulgated through joint consultative and other similar bodies concerned with health and safety that are present in it. Indeed, there would seem a case where subcontracting employers are to be present for reasonably lengthy periods, such as can occur on construction projects, to additionally require that they take part in the proceedings of these bodies, at least where the number of workers involved exceeds a certain threshold.

By itself, the mere creation of such legislative provisions will not resolve the difficulties that surround the representation of those working under non-standard forms of employment that stem from the differing nature of their material and psychological attachment to the organisations they are working for and the fact that their interests may in certain respects clash with those employed on a 'permanent' basis (James, 2004). For example, the latter may view the use of such staff as being undesirable and something that should be resisted, while the former may be hesitant to overtly engage in collective processes that could be seen to endanger current and future employment opportunities. It would, however, seem reasonable to argue that the introduction of legal provisions along the lines proposed would help in overcoming difficulties of this type by (a) legitimising the involvement of non-standard workers in such processes, particularly if suitable protections against victimisation are provided, (b) encouraging union and non-union representatives to focus more attention on addressing their concerns and (c) more generally, via these effects, strengthening the potential that exists for collective solidarity at the workplace level.

It does also need to be recognised that it may not be viable to establish representative arrangements in very small workplaces. There is no reason, however, for this potential problem to be used to support a move away from the principle that all employers should be covered by prescribed legal frameworks of representation and therefore to endorse the current use made of size thresholds. The Swedish use of roving safety representatives described by Kaj Frick in Chapter 8 serves to highlight the fact that an alternative approach is available. This is to provide small employers with the option, subject to their workforce's agreement, of placing reliance on such externally based trade union representatives.

The precise mechanisms through which access to representatives of this type could be provided would clearly need to be thought through. It is clear, however, that steps would need to be taken to enable unions to financially support the creation of a pool of such representatives, either through making available government funding or, perhaps more desirably, requiring employers to fund the scheme, perhaps on a 'consultancy fee' basis or through a national payroll levy system.

Mechanisms of enforcement

In an ideal world, perhaps, workers and their representatives would have the capacity, through a combination of their power and expertise, to take full advantage of the legal rights of representation that exist. Such a vision of 'self-regulated' implementation, however, is clearly fanciful in a context of weakened trade unions, more fragmented and casualised internal and external labour markets, heightened job insecurity arising from competitive product market pressures, and, more generally, resurgent employer power. It is, though, one that, from the expansion of representative rights in the 1970s and 1980s to the present day, continues, albeit often implicitly, to hold sway among policymakers. As a result, the issue of adequately *enforcing* the rights provided has, as this volume demonstrates, received a low priority. Instead, evidence of declining representative coverage has been generally met at best, as developments in Britain illustrate well, by policy pronouncements stressing the value of workforce representation, or at least the watered-down notion of 'workforce involvement', and calls upon employers to take steps to support and develop it (James and Walters, 2005).

The evidence provided in the chapters in this collection demonstrates clearly that this policy response has been insufficient to ensure the widespread implementation of legally mandated rights of representation. In fact, it has been associated not only with a very partial picture of implementation, even, as Theo Nichols and David Walters show in Chapter 1, in relatively favourable workplace environments, but one that has become less rather than more comprehensive over time in the face of the types of changes in the world of work mentioned in the previous paragraph.

Now more than ever, therefore, action is needed to substantially improve the enforcement mechanisms that exist to support the implementation of legal frameworks for worker representation. Such action, as the contributions to this volume have highlighted, would seem to require three main, and mutually supporting, types of reform: an expansion of government inspector numbers and hence inspections; greater prioritisation of the enforcement of representative rights; and a 'democratisation of enforcement' through empowering representatives and unions to themselves enforce the rights of representation granted to them.

As regards the first two of these issues, the evidence available on the current situation in Britain serves to graphically illustrate both the need for action and the scale of the problem that it needs to address. For example, the Health and Safety Executive (HSE) is the body responsible for enforcing health and safety laws in the most hazardous sectors of work in the British economy, but it has been estimated that a workplace in these sectors, on average, faces a likelihood of receiving an inspection once in every 13 years (House of Commons Work and Pensions Committee, 2008). Meanwhile, the paucity of enforcement notices and prosecutions brought against employers in respect of failures to carry out their duty to consult with workers or safety representatives, under respectively the Health and Safety (Consultation with Employers) Regulations 1996 and the Safety Representatives and Safety Committees Regulations 1977, or to comply more generally with the provisions of these regulations, highlights only too well the limited threat of enforcement action that employers face. This situation, it should additionally be noted, is unsurprising given that long-standing HSE policy advice to inspectors effectively requires that they consider enforcement action only after all other avenues for resolving any labour–management disputes over the issue of non-compliance have been unsuccessfully tried (James and Walters, 2005).

Yet, notwithstanding the need to increase inspector numbers, the fact remains that, inevitably, there will never be enough inspectors in any country to comprehensively monitor employer compliance with representative rights or health and safety laws more generally. An increase in inspector numbers, no matter how substantial, would in all likelihood therefore be insufficient in itself. It is for this reason that there would seem a case for taking the earlier mentioned philosophy of 'self-regulated implementation' further by extending it to encompass the empowering of workers and their representatives and unions to initiate enforcement action themselves.

One obvious way of doing this would be to borrow from the use made of the provisional improvement notice systems utilised in some parts of Australia. Another would be to allow, where the national legal system permits this to be done, workers and those that represent them to bring private prosecutions against employers in certain defined circumstances. In either case, however, it would seem desirable, given that it cannot be simply assumed that workers have the confidence or knowledge to pursue legal action against their employer, to buttress these powers with a union right of entry to workplaces where it is believed that employers are failing to comply with their legal obligations.

Linkages to wider employment relations structures

A number of the contributors to this volume have, with varying degrees of explicitness, drawn attention to how the operation of health and safety representative arrangements is influenced by the broader nature of employment relationships within which they are located. In doing so, they have therefore emphasised the way in which the latter can act to support (or undermine) the effectiveness of the former.

This last observation points, at a general level, to the value of locating health and safety representation within broader-based structures of worker representation. The case for this takes on more weight when it is borne in mind that (a) the most common forms of work-related ill health, at least in industrialised countries, namely, musculoskeletal disorders and stress-related disorders (see for example in the British case, HSE, 2007), are intimately connected to the way in which work tasks and processes are designed and operated, and (b), as Ana M. Garcia and colleagues in Chapter 4 show in Spain, it is precisely within this area that safety representatives most commonly experience difficulties in terms of exerting an influence.

The potential value of health and safety representation being located within a broader framework of workplace representation in effect receives an acknowledgement in France by virtue of the fact that, as Thomas Coutrot highlights in Chapter 3, health and safety committees are designated as 'subsections' of also legally mandated works councils. Elsewhere, such an explicit linkage of this type is generally absent. Admittedly, where unions are present, it may well be the case that such a linkage in practice exists, as, for example, will often be the case with safety representatives appointed under the British Safety Representatives and Safety Committees Regulations given that they are union-appointed. The fact remains, however, that in many workplaces, particularly non-unionised ones, health and safety representation, in the absence of legal provisions of the type found in France, is likely to be isolated from any more general and potentially supportive system of worker representation. Representation that is, perhaps, better placed to exert an influence over broader areas of employment relations, such as work organisation and working time, that can adversely affect worker health and more general well-being, and engender collective solidarity among workers.

Ideally, then, legislative provisions aimed at avoiding such an isolated form of health and safety representation would appear desirable. At present, the scope to do this is admittedly limited by the fact that many countries, such as Britain, Australia and Canada, lack legal frameworks which provide for the mandatory presence of a broader-based representative structure. Indeed, the situation in this regard is, internationally, generally worse than that which currently exists in relation to health and safety representation. This position, though, doesn't negate the virtue of seeking mutually supporting sets of legal frameworks relating to health and safety representation on the one hand, and wider employment representation on the other. Instead, it merely highlights the fact that action aimed at further developing the former cannot sensibly be pursued in isolation from the issue of worker representation more generally.

Such wider representative frameworks would, however, need to embody the same comprehensiveness of coverage as has been proposed above in relation to health and safety representation. It would also seem important for them to provide workers with rights of representation that offer real opportunities to exert an influence over issues of work organisation, including the design of tasks and processes, workloads, the length and distribution of working time, and job autonomy, that are intimately connected to psychosocial causes of work-related ill health (Anderson-Connolly et al., 2002; Taylor et al., 2003; James, 2006). Current difficulties that health and safety representatives face in influencing factors of this type reflect the fact that such factors are at the heart of managerial prerogatives concerning the governance of how work is organised and carried out. They therefore involve areas of decision-making where employers are likely to be strongly resistant to greater worker influence.

This last point, in turn, raises a challenge for unions. For it points to the fact that they need to ensure that health and safety is not treated as a distinct, 'technical', issue but as one that constitutes a key and integrated part of their more general representational and organising work: an integration which Rebecca Loudoun and Davis Walters suggest in Chapter 9 is currently underdeveloped.

Future prospects

The reforms identified above undoubtedly constitute a radical agenda for reform given that they would require significant changes to be made to the legal frameworks for health and safety representation that are currently in place internationally. They are also undoubtedly ambitious when set alongside prevailing social, political and economic contexts that have for the last three decades been marked by a decline in the power of organised labour and a rise of neo-liberal governmental philosophies, with all this has entailed for the lobbying influence wielded by employers and their organisations. This is particularly so when recent legislative developments in the area, as Kaj Frick in Chapter 8 and Laurent Vogel and David Walters in Chapter 5 demonstrate, have at times encompassed reverses rather than advances.

This ambition, though, needs to be itself set alongside the strong case that exists to support the identified reforms, as demonstrated by the contributions in this volume as well as other wider evidence. Thus, work does still impose enormous harm on the physical and mental health welfare of workers. There is firm evidence to indicate that worker representation on workplace health and safety matters can do much to reduce the level of this harm. And it is also clear that the capacity of current legal frameworks for health and safety representation to achieve such a reduction is hampered by weaknesses in both their provisions and the way in which these are currently implemented.

While the achievement of reform along the lines suggested is likely to be difficult, the case for it is a strong one that merits pursuing within individual countries and also at the level of the European Union. This is particularly so given that it cannot be simply assumed that even within the prevailing social, political and economic contexts 'windows of reform opportunity' won't appear. Who, for example, could have foreseen that a financial crisis would arise towards the end of the present decade which would prompt the United States and United Kingdom governments to embark on programmes of regulatory reform encompassing the effective nationalisation of banks and actions to reverse key aspects of the processes of financial liberalisation that had previously been pursued?

What is clear, however, is that, if such opportunities are to be seized, unions and other progressive political forces must not be afraid to embrace such an agenda of reform and to also pursue it rigorously.

References

AFA (2008) 080418: http://www.afaforsakring.se/WmTemplates/Page.aspx?id=200.

AFL-CIO (2008) 080416: http://www.aflcio.org/issues/safety/ergo/.

Almquist, R. and Henningsson, J. (2009) 'When capital market actors reduce the complexity of corporate personnel and work environment information,' *Journal of Human Resource Costing and Accounting*, 13(1): 46–60.

Altmann N. (1992) 'Unions' policies towards new technologies in the 1980s – an example from the metal industry' in: N. Altmann, C. Köhler and P. Meil (1992), *Technology and Work in German Industry*, London, Routledge: 361–85.

Amossé T., Bloch-London C. and Wolff L. dir. (2008) *Les relations sociales en entreprise – Un portrait à partir des enquêtes Relations Professionnelles et Négociations d'Entreprise (REPONSE 1992–3, 1998–9, 2004–5)*, Paris (ed.) La Découverte.

Anderson-Connolly, R., Grunberg, L., Greenberg, E. and Moore, S. (2002) 'Is Lean Mean? Workplace Transformation and Employee Well-Being', *Work, Employment and Society*, 16 (3): 389–413.

Antila, J. and Ylöstalo, P. (1999) *Working Life Barometer in the Baltic Countries 1998, Labour Policy Studies No. 214*, Helsinki, Ministry of Labour.

Antila, J. and Ylöstalo, P. (2003) *Working Life Barometer in the Baltic Countries 2002, Labour Policy Studies No. 247*, Helsinki, Ministry of Labour.

Antonsson, A. B., Birgersdotter, L. and Bornberger-Dankvardt, S. (2002) *Small enterprises in Sweden. Health and safety and the significance of intermediaries in preventive health and safety*, Arbete och hälsa 2002: 1, Stockholm, Arbetslivsinstitutet.

Antonsson, A. B. and Schmidt, L. (2003) *Småföretag och företagshälsovård – ska berget komma till Muhammed eller Muhammed till berget?*, IVL rapport B1542, Stockholm, IVL Svenska Miljöinsitutet.

Aronsson, G. (1999) *Contingent Workers and Health and Safety, Work Employment Society* 13: 439–59.

Aronsson, G. and Gustafsson, K. (1999) *Kritik eller tystnad – en studie av arbetsmarknads- och anställningsförhållandens betydelse for arbetsmiljökritik*, Arbete och Hälsa 2005: 5, Arbetslivsinstitutet.

Aronsson, G., Gustafsson, K. and Dallner, M. (2002) 'Work environment and health in different types of temporary jobs', *European Journal of Work and Organizational Psychology*, 11 (2): 151–75.

ACTU (2002) *A Report on the 2001 National Survey of Health and Safety Representatives*, Australian Council of Trade Unions, Melbourne, September.

ACTU (2005) *A Report on the 2004 National Survey of Health and Safety Representatives*, Australian Council of Trade Unions, Melbourne, November.

AV (2004) *Årsredovisning 2003*, Solna, Arbetsmiljöverket.

AV (2006) *Rapport 2006: 5*, Frick, K., Bruhn, A. and Lehto, A. 'Metodutveckling. En delrapport i utvärderingen av utvecklingsprogrammet ARbetsorganisation och NEgativ stress (ARNE)', Solna, Arbetsmiljöverket.

AV (2007) *Sammanställning över inkomna redovisningar*. ADVUS 2005/55982 rapport 18 June 2007, Solna, Arbetsmiljöverket.

AV (2008a) *Arbetsmiljöverkets Budgetunderlag 2009–11*. LS 2008/5215, 27 February 2008, Solna, Arbetsmiljöverket.
AV (2008b) *Årsredovisning 2007*, Solna, Arbetsmiljöverket.
AV (2008c) 080318: dodsfall_utlandsk_2005-2007.pdf, Solna, Arbetsmiljöverket.
Axelsson, C. (2002) *Formalisering som hinder och möjlighet i små företag*, Rapport 2002: 14, Doctoral Thesis, Luleå, Luleå Tekniska Universitet.
Bagnara, S., Misiti, R. and Wintersberger, H. (eds) (1985) *Work and health in the 1980s: experiences of direct worker's participation in occupational health*, Berlin, Sigma.
Barrefeldt, B. (2008) Telephone communication, Swedish Ministry of Labour, May 19.
Barry, M. and Loudoun, R. (2006) 'Industrial relations, occupational safety and health and union organising in Australia: lessons and opportunities', *Policy and Practice in Health and Safety*, 4 (1): 31–44.
Beck, M. and Woolfson C. (2000) 'The regulation of health and safety in Britain: From old labour to new labour', *Industrial Relations Journal*, 31 (1): 35–50.
Beck, M., Robson, M., Watterson, A. and Woolfson, C. (2002) Exploring health, safety and environment in central and eastern Europe, *New Solutions: Journal of Occupational and Environmental Health Policy*, 11 (3): 6–10.
Bernstein, S., Lippel, K., Tucker, E. and Vosko, L. (2006), 'Precarious employment and the law's flaws: Identifying regulatory failure and securing effective protection for workers' in L. Vosko (ed.) *Precarious Employment: Understanding Labour Market Insecurity in Canada*, Montreal, McGill Queen's University Press: 203–20.
Biggins, D. and Phillips, M. (1991a) 'A survey of health and safety representatives in Queensland Part 1: Activities, issues, information sources', *Journal of Occupational health and Safety – Australia and New Zealand*, 7 (3): 195–202.
Biggins, D. and Phillips, M. (1991b) 'A survey of health and safety representatives in Queensland Part 2: Comparison of representatives and shop stewards', *Journal of Occupational Health and Safety – Australia and New Zealand*, 7 (4): 281–6.
Biggins, D. and Holland, T. (1995) 'The training and effectiveness of health and safety representatives' in I. Eddington, *Towards Health and Safety at Work: Technical Papers of the Asia Pacific Conference on Occupational Health and Safety*, Brisbane.
Blewett, V. (2001) *Working together: A review of the effectiveness of the health and safety representative and workplace health and safety committee system in south Australia – Final Report and Recommendations*, Consultative Arrangements Working Party, Adelaide, WorkCover Corporation of South Australia.
Bos, N. (1995) 'Health and safety representatives: Contributing to improved health and safety in the workplace' in I. Eddington (ed.) *Towards Health and Safety at Work: Technical Papers of the Asia Pacific Conference on Occupational Health and Safety*, Brisbane.
Brooks, A. (1993) *Occupational health and safety law in Australia*, Sydney, CCH Limited, 3rd edition.
Bruun, N., Flodgren, B., Halvorsen, M. and Hydén, H. (1992) *The Nordic labour relations model: Labour law and trade unions in the Nordic countries – today and tomorrow*, Aldershot, Dartmouth.

Carcoba, A. (2007) *La salud no se vende ni se delega, se defiende: el modelo obrero*, Madrid (ed.) GPS.

Carnevale, F. (2007) La santé des travailleurs en Italie: le rôle décisif des luttes ouvrières, *Histoire et Sociétés*; 23: 44–55.

Cézard M., Malan A. and Zouary P. (1996) Conflitetrégulationsocialedans les établissements, *Travail et Emploi*, 66: pp. 19–38

Clarke, M., Lewchuk, W., de Wolff, A. and King, A. (2007) '"This just isn't sustainable"': precarious employment, stress and workers' health' *International Journal of Law and Psychiatry*, 30 (6): 311–26

Commission for Labor Cooperation (2003) Recent trends in union density in north America. Washington, http://www.naalc.org/english/pdf/april_03_english.pdf.

Coutrot, T. (2002) *Critique de l'organisation du travail*, Paris, La Découverte.

Coutrot, T. (2008) La prévention des risques professionnels vue par les salariés, *Premières Synthèses* 5 (1): 1–8.

Creighton, B. and Rozen, P. (2007) *Occupational health and safety law in Victoria*, Sydney, Federation Press, 3rd edition.

Culvenor, J., Cowley, S. and Harvey, J. (2003) 'Impact of health and safety representative training on concepts of accident causation and prevention' *Journal of Occupational Health and Safety Australia and New Zealand*, 19 (3): 279–92.

Dares (2007) 'Eléments de cadrage statistique', *Conférence tripartite sur l'amélioration des conditions de travail*, Paris, Ministère du travail.

Davies, P. and Kilpatrick, C. (2004) 'UK worker representation after single channel', *Industrial Law Review*, 33 (2): 121–51

Davis, C (2004) *Making Companies Safe: What Works?* London, Centre for Corporate Accountability.

Didry (2001) 'Le Comité d'entreprise européen devant la justice: mobilisation du droit et travail juridique communautaire', *Revue Droit et Société*, 49: 911–34.

Digby, C. and Riddell C. (1986) 'Occupational Health and Safety in Canada' in *Canadian Labour Relations*, Volume 16, Royal Commission on the Economic Union and Development Prospects for Canada, Craig Riddell (ed.), Toronto: University of Toronto Press: 285–320.

Direktiv (2007) *Utredningsdirektiv 2007: 116*, Stockholm, Arbetsmarknadsdepartementet.

Direktiv (2008) *Utredningsdirektiv 2008: 38*, Stockholm, Arbetsmarknadsdepartementet.

Dølvik, J. E. and Eldring, L. (2006) *The Nordic Labour Market two years after the EU enlargement – Mobility, effects and challenges*, TemaNord, 2006: 558, Copenhagen, Nordic Council of Ministers.

Ds (2008) *Arbetsmiljön och utanförskapet*, Ds 2008: 16, Stockholm, Arbetsmarknadsdepartementet.

DTI (2004) *Regulations to establish a general framework for informing and consulting employees in the UK: Full regulatory assessment*, London, DTI, Employment Relations Directorate, October http://www.dti.gov.uk/er.

Elling, R. (1986) *The struggle for workers' health: A study of six industrialized countries*, Farmingdale, Baywood.

Estabrook, T. (2007) *Labor-Environmental Coalitions: Lessons from a Louisiana Petrochemical Region*, Amityville, New York, Baywood.

European Agency (2002) *Promoting health and safety in European small and medium-sized enterprises (SMEs)*. Bilbao, European Agency for Safety and Health at Work.

European Commission (1985) *White Paper 1985*, COM (85) 310 final, Brussels: European Commission.

European Commission (1989) Council Directive 89/391/EEC of 12 June 1989 on the introduction of measures to encourage improvements in the safety and health of workers at work. Official Journal. L 183, 29/06/1989 P. 0001 – 0008. Brussels: European Commission. http://eurlex.europa.eu/LexUriServ/ LexUriServ.do?uri=CELEX:31989L0391:EN:HTML.

European Commission (2002) *Adapting to change in work and society: A new Community strategy on health and safety at work 2002–6.* Communication from the Commission.COM (2002) 118 final. Brussels: European Commission. http://ec.europa.eu/employment_social/news/2002/mar/new_strategy_en.pdf.

European Commission (2004) Communication on the practical implementation of the provisions of the Health and Safety at Work Directives: 89/391 (Framework), 89/654 (Workplaces), 89/655 (Work Equipment), 89/656 (Personal Protective Equipment), 90/269 (Manual Handling of Loads) and 90/270 (Display Screen Equipment), COM (2004) 62, final, Brussels: European Commission.

European Commission (2007) Communication from the Commission to the Council and the European Parliament. *Improving Quality and Productivity at Work: Community strategy 2007–12 on health and safety at work.* Brussels, 21 February, COM (2007) 62 Brussels: European Commission. http://ec.europa. eu/employment_social/news/2007/feb/commstrat_en.pdf.

European Foundation for the Improvement of Living and Working Conditions (2002) *First candidate countries survey on working conditions (2001)* questionnaire, Dublin, European Foundation for the Improvement of Living and Working Conditions. http://www.eurofound.europa.eu/working/surveys/documents/ questionnaire_en-cc12.pdf.

European Foundation for the Improvement of Living and Working Conditions *EIRO* (2005a) *Lithuania-Trade unions in focus*, Dublin, European Foundation for the Improvement of Living and Working Conditions. http://www.eiro. eurofound.eu.int/2004/12/feature/lt0412102f.html.

European Foundation for the Improvement of Living and Working Conditions *EIRO* (2005b) *Latvia – Trade unions seek to boost membership*, Dublin, European Foundation for the Improvement of Living and Working Conditions. http:// www.eiro.eurofound.eu.int/2005/12/feature/lv0512102f.html.

European Foundation for the Improvement of Living and Working Conditions *EIRO* (2006) *Estonia-Survey examines social partnership in enterprises*, Dublin, European Foundation for the Improvement of Living and Working Conditions. http://www.eiro.eurofound.eu.int/2006/02/feature/ee0602102f.html.

European Foundation for the Improvement of Living and Working Conditions (2007) Press Pack. '% considers their work affects their health', Dublin, European Foundation for the Improvement of Living and Working Conditions. http:// www.eurofound.europa.eu/docs/press/ewcs2005/8workrelatedhealthoutcomes.xls.

Eurostat (2006) *Structural Indicators* http://epp.eurostat.ec.europa.eu/portal/ page?_pageid=1996,39140985and_dad=portaland_schema=PORTALandscreen= detailrefandlanguage=enandproduct=Yearlies_new_economyandroot=Yearlies_ new_economy/B/B1/B11/eb021.

Ewing, K. D. and Truter, G. M. (2005) 'The information and consultation of employees regulations: voluntarism's bitter legacy', *Modern Law Review*, 68 (4): 626–41.

Fairbrother, P. (2002) 'Unions in Britain: Towards a new unionism' in P. Fairbrother and G. Griffin (eds), *Changing prospects for unionism: Comparisons between Six Countries*. London, Continuum: 56–93.

Fine, J. (2005) *Workers centers: Organising communities at the edge of the dream*, Washington DC: Economic Policy Institute Briefing Paper.

Frick, K. (1979) *Skyddsarbetet i små industriföretag*, Stockholm: Arbetarskyddsnämnden.

Frick, K. (1995) *Reconciling the divergent goals of high productivity and occupational health and safety – The Swedish Work Life Funds*, Working Paper, Stockholm, National Institute for Working Life.

Frick, K. (1996) *De regionala skyddsombudens verksamhet*, Rapport 1996: 22, Stockholm, Arbetslivsinstitutet.

Frick, K. (2002) 'Sweden: Occupational health and safety management strategies from 1970 – 2001' in D. Walters (ed.), *Regulating Health and Safety Management in the European Union*, Brussels: P.I.E. Peter Lang.

Frick, K. (2005) 'Inblick: Blir privata tjänster nästa ohälsosektor?', *Tidningen Arbetarskydd* 2005 (6): 12.

Frick, K. (2008) *Experiences and possibilities to develop psychosocial health in small firms*, Half-day seminar 1 April 2008 with 39 Regional safety representatives from 10 LO unions, Västerås: Mälardalen University.

Frick, K., Eriksson, O. and Westerholm. P. (2005) 'Work environment policy and the actors involved' in R. Å. Gustafsson and I. Lundblad (eds), *Worklife and Health in Sweden 2004*, Stockholm, National Institute for Working Life.

Frick, K., Jensen, P., Quinlan, M. and Wilthagen, T. (eds) (2000), *Systematic occupational health and safety management – Perspectives on an international development*, Oxford Pergamon.

Frick, K., Sigala, F. and Sundström-Frisk, C. (1997) *LOs regionala skyddsombud: Skillnader från 1980 till 1993 samt mellan förbunden*, Rapport 1997: 20, Solna, Arbetslivsinstitutet.

Frick, K. and Sjögren, D. (1980) *Förebyggande eller i förbigående – skyddsarbetet i 18 små industriföretag*, Stockholm, Arbetslivscentrum.

Frick, K. and Sjöström, J. *Factors influencing worker and safety representative participation – How to understand the OHS participation process* (accessed 1 May 2006), http://hesa.etui-rehs.org/uk/dossiers/dossier.asp?dos_pk=15.

Frick, K. and Walters, D. (1998) 'Worker representation on health and safety in small enterprises: Lessons from a Swedish approach', *International Labour Review*, 137 (3): 395–417.

Fudge, J., Tucker, E. and Vosko, L. (2002), *The legal concept of employment: Marginalizing workers, report for the Law Commission of Canada*: 1–141.

Gall, G. (2003) *Union organising: Campaigning for Trade Union Recognition*, London, Routledge.

Gall, G. (2006) *Union recognition. Organising and bargaining outcomes*, London, Routledge.

Gallagher, C. (2001) 'New directions: innovative management plus safe place' in W. Pearse, C. Gallagher and E. Bluff (eds), *Occupational health and safety management systems, Proceedings of the first national conference*, University of Western Sydney, July 2000, Crown Content, Melbourne, 65–82.

Gallagher, C., Underhill, E. and Rimmer, M. (2003) 'Occupational health and safety management systems in Australia: Barriers to success' *Policy and Practice in Health and Safety* 1 (2): 67–81.

Garcia, A. M., Boix, P. and Canosa, C. (2004) 'Why do workers behave unsafely at work? Determinants of safe work practices in industrial workers'. *Occupational and Environmental Medicine*, 61: 239–46.

García, A. M., Gadea, R. and Rodrigo, F. (2005) 'Occupational risk prevention in workplaces: Perceptions of health and safety representatives in Spain'. *Archivos de Prevención de RiesgosLaborales*, 8: 139–46 (in Spanish).

Garcia, A. M., Lopez-Jacob, A. M., Dudzinski, I., Gadea, R. and Rodrigo, F. (2007) 'Factors associated with the activities of safety representatives in Spanish workplaces', *Journal of Epidemiology and Community Health*, 61: 784–90.

Geldart, S., Shannon, H. and Lohfield, l. (2005) 'Have companies improved their health and safety approaches over the last decade? A longitudinal study', *American Journal of Industrial Medicine*, 47: 227–36.

Gellerstedt, S. (2007) *Samverkan för bättre arbetsmiljö – skyddsombudens arbete och erfarenheter*, Stockholm: Landsorganisationen.

Gollac, M. (1998) *Donner unsens aux données. L'exemple des enquêtesstatistiquessur les conditions de travail*, Cahiers du CEE, Documentation Française

Gospel, H. and Wood, S. (2003) *Representing Workers*, London, Routledge.

Grayson, J. and Goddard, C. (1975) *Industrial Safety and the Trade Union Movement, Studies for Trade Unionists*, 1 (4) .

GS (Goldman Sachs) Economic Website (2007) 'Baltic Boom: The adjustment is likely to be painful and to start soon', *Global Viewpoint*, 7 (30), 6 November.

Gunningham, N. and Johnstone, R. (1999) *Regulating Workplace Safety: Systems and Sanctions*, Oxford, Oxford University Press.

Gunningham, N. (1985) 'Workplace Safety and the Law' in W. B. Creighton and N. Gunningham (eds) *The industrial Relations of Occupational Health and Safety*, Sydney, Croom Helm.

Gustavsen, B. (2007) 'Work Organization and "the Scandinavian Model"', *Economic and Industrial Democracy*, 28 (4): 650–71.

Hale, A. and Hovden, J. (1998) 'Management and culture: The third age of safety. A review of approaches to organisational aspects of safety, health and environment', in A. Feyer and A. Williamson (eds), *Occupational injury. Risk prevention and intervention*, London, Taylor and Francis: 129–66.

Hall, A., Forrest, A. Sears, A. and Carlan, N. (2006) 'Making a difference: Knowledge activism and worker representation in Joint OHS Committees', *Relations Industrielles/Industrial Relations*, 61 (3): 408–36.

HSE (2007) *Health and safety statistics 2006/07*. Sudbury: HSE Books.

Heery, E. and Simms, M. (2008) 'Constraints on union organising', *Industrial Relatons Journal*, 39: 24–42.

Hillage, J., Kersley, B., Bates, P. and Rick, J. (2000) *Workplace consultation on health and safety*, HSE Contract Research Report 268/2000, London: HSE Books.

Holgate, J. and Simms, M. (2008) 'TUC organising Academy 10 Years on: What has been the impact?' paper presented at Centre for Global Labour Research seminar, Cardiff University, 17 April 2008.

Holmgren, A. (2008) *RITA – rättvis ingång till arbete. En rapport om papperslösa i städbranschen*, Stockholm: Fastighetsanställdas förbund.

House of Commons Work and Pensions Committee (2008) *The role of the health and safety commission and the health and safety executive in regulating workplace health and safety: Volume 1.* London: Stationary Office.

Human Resources and Social Development Canada (2007), *Union membership in Canada – 2007*, http://www.hrsdc.gc.ca/en/lp/wid/union_membership. shtml.

Hutton, J. (2008) *25 ideas for simplifying EU law*, London, Better Regulation Executive.

Hyman, R. and Ferner, A. (1994) *New frontiers in European Industrial Relations*, London, Blackwell.

Instituto per il Lavoro (2006) 'The role of the safety representative in Italy', http://hesa.etui-rehs.org/ukdossiers/files/IPL.pdf.

International Labour Office (2006), *Labour inspection audit: Tripartite audit of the labour inspection system of Latvia 3–14 October 200*, Geneva, International Labour Office.

ISE (Institut Syndical Européen) (1982) *Négocier l'introduction de nouvelles technologies. Aperçu des différentes approches adoptées par les syndicats d'Europe occidentale à l'égard de l'introduction des nouvelles technologies*, Brussels, ETUI.

ISE (Institut Syndical Européen) (1985), *Nouvelles technologies et négociations collectives. Le bilan de dix années d'expérience en Europe*, Brussels, ETUI.

IVA/NUTEK (2006) *Internationaliseringen driver på strukturomvandlingen*, Nya fakta och statistik no. 9, Stockholm: Ingenjörsvetenskapsakademin-NUTEK.

James, P. and Walters, D. (1997) 'Non-union rights of involvement: The case of health and safety at work', *Industrial Law Journal* 26: 35–50.

James, P. (2004) 'Trade unions and non-standard workers' in M. Harcourt and G. Wood (eds), *Trade unions and democracy: Strategies and perspectives*, Manchester: Manchester University Press: 82–104.

James, P. (2006) 'The changing shape of workplace risk' in P. James and G. Wood (eds), *Institutions, Production and Working Life*, Oxford: Oxford University Press: 255–72.

James, P., Johnstone, R., Quinlan, M. and Walters, D. (2007) 'Regulating supply chains to improve health and safety', *Industrial Law Journal*, 36 (2), 163–87.

James, P. and Walters D. (eds) (1999), *Regulating health and safety at work: The way forward*, London, Institute of Employment Rights.

James, P. and Walters, D. (2005) *Regulating health and safety at work: An agenda for change*, London, Institute of Employment Rights.

Jensen, P. L. (2002) 'Assessing assessment: The Danish experience of worker participation in risk assessment', *Economic and Industrial Democracy*, 23 (2): 201–227.

Jensen, C. (2005) 'European Trade Unions: Influence and members – a comparative analysis of workplace influence and member composition among trade unions in Europe', *Paper presented at the annual meeting of the American Sociological Association*, Philadelphia.

Johansson, B. (1998) *En motivationsbaserad analys av arbetsmiljöarbetet i en grupp tillverkande mindre företag*, Rapport 1998: 36, Doctoral Thesis, Luleå: Luleå Tekniska Högskola.

Johansson, K. (2007) Telephone interview with the OHS-officer of the Swedish construction workers union, December 7.

Johnstone, R. (2004) *Occupational health and safety law and policy: Text and materials*, Sydney, LBC Information Services, 2nd edition.

Johnstone, R., Quinlan, M. and Walters, D. (2005) 'Statutory OHS workplace arrangements for the modern labour market', *Journal of Industrial Relations* 47: 93–116.

Kelly, J. and Frege, C. M (eds) (2004) *Varieties of unionism: Strategies for union revitalization in a globalizing economy*, Oxford, Oxford University Press.

Kersley, B., Carmen, A., Forth, J., Bryson, A., Bewley, H., Dix, G. and Oxenbridge, S. (2006) *Inside the workplace: Findings of the 2004 workplace employment relations survey*, London, Routledge.

Knowles, D. (2006) *Measuring the effect of health and safety advisers and roving safety representatives in agriculture*, Norwich, Health and Safety Executive Research Report RR 417 prepared by ADAS Consulting Ltd, HMSO, Online at http://www.hse.gov.uk.

Knudsen, H. (1995) *Employee participation in Europe*, London, Sage.

Kohl, H. and Platzer, H-W. (2004) *Industrial relations in Central and Eastern Europe* Brussels, European Trade Union Institute.

Lax, M. B. (2006) 'Inspiration for a movement: Re-reading death on the job', *New Solutions*, 16 (3) 315–48.

Leacy, F. H. (1983), *Historical statistics of Canada*, Series D236-259, http://www.statcan.ca/english/freepub/11-516-XIE/sectiona/toc.htm.

Lee, M. and Quinlan, M. (1994) 'Co-ordinating the regulation of occupational health and safety in Australia: Legal and political obstacles and recent developments' *Australian Journal of Labour Law*, 7 (33): 1–35.

Levinson, K. (2004) *Lokal partssamverkan – en undersökning av svenskt medbestämmande*, Arbetsliv i omvandling, 2004: 5, Stockholm: Arbetslivsinstitutet.

Levesque, C. (1995), 'State Intervention in Occupational Health and Safety: Labour-Management Committees Revisted' in A. Gilesand K. Wetzel (eds), *Proceedings of the XXXIst Conference of the Canadian Industrial Relations Association*, Toronto: 217–31.

Lewchuk, W., Robb, L. and Walters, V. (1996). 'The effectiveness of Bill 70 and Joint Health and Safety Committees in reducing injuries in the workplace: The case of Ontario' *Canadian Public Policy*, 23: 225–43.

Lewchuk, W., Clarke, M. and de Wolff, A. (2008) 'Working without commitments: Precarious employment and health', *Work Employment and Society*, 22 (3): 387–406.

Lippel, K. (2006) 'Precarious employment and occupational health and safety regulation in Quebec' in L. Vosko (ed.), *Precarious Employment: Understanding Labour Market Insecurity in Canada*, Montreal, McGill Queen's University Press: 241–55.

Littorin, S. O. (2008) 080407: 'Vi värnar om den svenska modellen', *Nya Wermlands-Tidningen*.

LO (2004) *Arbetskraft till salu – sex månader med öppna gränser*, Stockholm: Landsorganisationen.

LO-tidningen (2006a) 'Utan fast kneg – trots 30 år i branschen', *LO-tidningen*, 21 December 2006: 10.

LO-tidningen (2006b) 'Industrin sparkar 11 000 och hyr in 11 000 i stället', *LO-tidningen*, 21 December 2006: 11.

LO-tidningen (2008a) 'Medlemmarna flyr. Facken tappar medlemmar i hela världen', *LO-tidningen*, 28 March 2008: 5.

LO-tidningen (2008b) 'Tappet fortsätter', *LO-tidningen*, 8 August 2008: 12.

Lysgaard, S. (1967) *Arbeiderkollektivet*, Oslo: Universitetsforlaget.

Maxwell, C (2004) *Occupational health and safety act review*, Melbourne, State of Victoria.

Mayhew, C., Young, C., Ferris, R. and Harnett, C. (1997) *An evaluation of the impact of targeted interventions on the OHS behaviours of small business building industry owners/managers/contractors*, Sydney, National Occupational Health and Safety Commission.

Meardi, G. (2007), 'More voice after more exit? Unstable industrial relations in central eastern Europe', *Industrial Relations Journal*, 38 (6): 503–26.

Menendez, M, Benach, J. and Vogel, L. (2008) *Draft report on the impact of safety representatives on occupational health: A European perspective* (The EPSARE project) Brussels, ETUI-REHS.

Migrationsverket (2008) *Beviljade uppehållstillstånd 1980–2007; Asylsökande till Sverige under 1984–2007; Antalet uppehållstillstånd enligt EES-avtalet 1994–2007*, 080428, www.migrationsverket.se.

Milgate, N., Innes, E. and O' Loughin, K. (2002) 'Examining the effectiveness of health and safety committees and representatives: A review'. *Work* 19: 281–90.

Milkman, R. and Voss, K. (eds) (2004) *Rebuilding Labour: Organising and organisers in the new union movement*, Cornell, ILR.

Millward, N., Bryson, A. and Forth, J. (2000) *All change at work: British employment relations 1980–98, As portrayed by the workplace industrial relations survey series*, London, Routledge.

Moorhead, A., Steele, M., Alexander, M., Stephen K. and Duffin, L. (1997) *Changes at Work: The 1995 Australian Workplace Industrial Relations Survey*, Melbourne, Longman.

Neal, A. (1998) 'Regulating health and safety at work: Developing European Union Policy for the Millennium' *International Journal of Comparative Labour Law and Industrial Relations*, 14 (3): 217.

Nelkin, D. (ed.) (1985) *The language of risk – Conflicting perspectives on occupational health*, Beverly Hills: Sage.

Nielsen, R. and Szyszczak, E. (1997) *The social dimension of the European Union*, Copenhagen, HandelshojskensForlag.

Nichols, T. (1986) 'Industrial injuries in British manufacturing in the 1980s', *Sociological Review*, 34 (2): 290–306.

Nichols, T. (1997) *The sociology of industrial injury*: London, Mansell.

Nichols, T. and Armstrong, P. (1973) *Safety or profit: Industrial accidents and the conventional wisdom*, Bristol: Falling Wall Press.

Nichols, T., Dennis, A. and Guy, W. (1995) 'Size of employment unit, and industrial injury rates in British manufacturing', *Industrial Relations Journal*, 26: 45–56.

Nichols, T., Walters, D. and Tasiran, A. C. (2007) 'Trade unions, institutional mediation and industrial safety: Some evidence from the UK', *Journal of Industrial Relations*, 49 (2): 211–26.

Nilsson, N. T. and Carlsson, S. (1979) *Anställningstrygghet i mindre och medelstora företag – En studie av sambandet mellan företags storlek och personalomsättning*, Stockholm, Arbetslivscentrum.

O'Grady, J. (2000) 'Joint health and safety committees: Finding a balance' in T. Sullivan (ed.) *Injury and the New World of Work*, Vancouver, UBC Press: 162–97.

Obach, B. K. (2004) *Labor and the environmental movement: The quest for common ground*, Cambridge, Massachusetts, MIT Press.

Paoli, P. and Merllié, D. (2000) *Third European Survey on working conditions*, Dublin, European Foundation for the Improvement of Living and Working Conditions. http://www.eurofound.europa.eu/pubdocs/2001/21/en/1/ef0121en.pdf.

Parent-Thirion., A., FernándezMacías, E., Hurley, J. and Vermeylen, G. (2007) *Fourth European working conditions Survey*, Dublin, European Foundation for the Improvement of Living and Working Conditions. http://www.eurofound.europa.eu/pubdocs/2006/98/en/2/ef0698en.pdf.

Parsons, M. I. (1988) 'Worker participation in Occupational Health and Safety: Lessons from the Canadian Experience', *Labor Studies Journal*, 13: 22–32.

Pavlinek, P. (2002) 'The role of foreign direct investment in the privatisation and restructuring of the Czech motor industry', *Post-communist Economies*, 14 (3): 359–79.

Persson, K. I. (1979) *Working conditions in small industries – English summary of research*, Stockholm, Arbetslivscentrum.

Quinlan, M. (2000) 'Precarious employment, work re-organization and the fracturing of OHS management' in K. Frick, P. Jensen, M. Quinlan T. Wilthagen, (eds), *Systematic Occupational Health and Safety Management: Perspectives on an International Development*, Oxford: Pergamon: 175–98.

Quinlan, M. and Mahew, C. (1999) 'Precarious employment and workers' compensation', *International Journal of Law and Psychiatry*, 5–6: 491–520.

Quinlan, M., Mayhew, C. and Bohle, P. (2001) 'The global expansion of precarious employment, work disorganisation and occupational health: A review of recent research', *International Journal of Health Services*, 31, 335–414.

Reilly, B., Paci, P. and Holl, P. (1995) 'Unions, safety committees and workplace injuries', *British Journal of Industrial Relations*, 33 (2): 273–88.

Republic of Lithuania (2005) *State labour inspectorate annual report 2005*, Vilnius, Ministry of Labour and Social Affairs.

Robens, Lord. (1972) *Safety and health at work: Report of the committee 1970–72*, Cmnd 5034, London, HMSO.

Rueda, S. (2004) 'Work-related injuries and trade union strength in the OECD'. *Archivos de Prevención de RiesgosLaborales*, 7: 146–52 (in Spanish).

SCB (2003) *Arbetsmiljön i småföretag*, Information om utbildning och arbetsmarknad 2003: 1, Stockholm and Solna, SCB and Arbetsmiljöverket.

SCB (2008) *Företag efter näringsgren SNI92, storleksklass och tid; Antal anställda efter näringsgren och sektor*, http://www.scb.se/templates/tableOrChart_20543.asp.

Schaerström, A. (2006) *Migration and work-related health – a tentative literature review*, Working Paper, Stockholm: National Institute for Working Life.

Shannon, H. S., Waiters, V., Lewchuk, W., Richardson, J., Moran, L. A., Haines, T. and Verma, D. (1996), Workplace organizational correlates of lost-time accident rates in manufacturing, *American Journal of Industrial Medicine* 29: 258–268.

Shaw, A., Blewett, V., Gunningham, N., Johnstone, R. and Baker-Goldsmith, H. (2007) *Review of the NT Work Health Act and Mining Management Act: Final Report.*

Shaw, N. and Turner R. (2003). *The Workers Safety Advisors (WSA) pilot.* Prepared by York Consulting, Health and Safety Executive, Research report 144, Sudbury, HSE Books.

Simard, M., Carpentier-Roy, M. C., Marchand, A. and Ouellet, F. (1999) *Organisational and psychosocial factors favouring the participation of workers in health and safety activities*, Institut de Recherche en Santé et en Sécurité du Travail, Rapport R-211: http://www.irsst.qc.ca/fr/_publicationirsst_662.html (in French).

SKL (2008) *Ökade socialbidragskostnader med Försäkringskassans nya policys*, Press Release, 2 May 2008, Stockholm: Sveriges Kommuner och Landsting.

Smismans, S. (2003) 'Towards a new Community strategy on health and safety at work', *International Journal of Comparative Labour Law and Industrial Relations*, 19 (1): 55–84.

Smith, D. (2000) *Consulted to death*, Winnipeg, Arbeiter Ring Publishing.

SOU (1972) *Bättre arbetsmiljö – delbetänkande avgivet av arbetsmiljöutredningen*, SOU 1972: 86, Stockholm, Arbetsmarknadsdepartementet.

SOU (1990) *Arbete och hälsa – betänkande av arbetsmiljökommissionen*, SOU 1990: 49, Stockholm, Arbetsmarknadsdepartementet.

SOU (2007) *Bättre arbetsmiljöregler II. Slutbetänkande av utredningen om Arbetsmiljölagen*, SOU 2007: 43, Stockholm, Arbetsmarknadsdepartementet.

Stanzani, C. and Kempa, V. (2004) *National arrangements for workers' health and safety representatives: Transposition and implementation of the Framework Directive*, TUTB Newsletter, 22–3: 57–60.

Statistics Canada (2007), *National income and expenditure accounts, quarterly estimates*, 13-001-XIB. 55 (4).

Statistics Canada (2008) *Labour Force Information*, 71-001-XIE, http://www.statcan.ca/bsolc/english/bsolc?catno=71-001-XIE.

Storey, R. (2005) 'Activism and the making of occupational health and safety law in Ontario, 1960s–1980' *Policy and Practice in Health and Safety*, 3 (1): 41–68.

Storey, R. (2009) 'Don't work too hard: Health and safety and workers' compensation in Canada' in B. Bolaria and H. Dickinson, (eds), *Health, Illness and Health Care in Canada, 4th Edition*, Toronto: Nelson Education Limited: 388–411.

Storey, R. and Tucker, E. (2006) 'All that is solid melts into air: Worker participation and occupational health and safety regulation in Ontario, 1970–2000' in V. Morgensen (ed.), *Worker Safety Under Siege: Labor, Capital, and the Politics of Workplace Safety in a Deregulated World*. Armonk, M. E. Sharpe, Inc.: 157–86.

Sweeney Research (2005) *KPI Survey Study No. 14374*, Melbourne, WorkSafe Victoria.

Szyszcak, E. (2000) *EC Labour Law*, Harlow, Longman/Pearson.

SZW (2005) 'A social Europe', Conference workshop documentation, Ministerie van SocialeZaken en Werkgelegenheid, November 8–9th.

Taylor, P., Baldry, C., Bain, P. and Ellis, V (2003) 'A unique working environment: Health, sickness and absence management in UK call centres', *Work, Employment and Society*, 17 (3): 435–58.

Torehov, C., Sigala, F., Sundstrom-Frisk, C. and Frick, K. (1996) *De regionala skyddsombudens verksamhet – deskriptiva data från en enkätundersokning*, Rapport 1996: 23, Solna, Arbetslivsinstitutet.

Trentin, B. (2001) 'Revisiter le contrôle ouvrier', *Cahiers Marxistes*, 218, 203–24.

TUC (Trades Union Congress) (2008) Organising for Health and Safety: A Workplace Resource, London, TUC.

Tucker, E. (2003) 'Diverging trends in worker health and safety protection and participation in Canada, 1985–2000', *Relations Industrielles/Industrial Relations*, 58: 395–426.

Tucker, E. (2007) 'Remapping worker citizenship in contemporary occupational health and safety regimes', *International Journal of Health Services*, 37: 145–70.

Tuohy, C. and Simard, H. (1993) 'The impact of JHSC in Ontario and Quebec', a study prepared for the Canadian Association of Administrators of Labour Law.

TUTB (1995) 'Molitor Group: Deregulation assault on health and safety', *TUTB Newsletter*, 1 October.

UGT (2008) 'Presentations on territorial safety representatives: Asturias and Emilia-Romagna' by Ana Isabel Álvarez Antón, UGT; and by Gianni Pedrazzini, CISL at *III Congreso de Salud Laboral de la UGT del País Valencia, 31 January – 1 February 2008*, Castellón, Spain.

Vanderkruk, R. (2003) 'Workplace health and safety representatives: The Queensland experience', *Journal of Occupational Health and Safety – Australia and New Zealand* 19 (5): 427–35.

Vaughan-Whitehead, D. (ed.) (2005) *Working and employment conditions in the new EU member states: Convergence or diversity?* Geneva, ILO-EC.

Verheugen, G. (2008) 'Reducing red tape for Europe', Speech at EU Conference *Cutting Red Tape for Europe*, June 20th, Brussels.

Victorian Trades Hall Council (VTHC) (2004) *The view from the frontline: A report on the experience of OHS reps*, Melbourne, Victorian Trades Hall Council.

Victorian Trades Hall Council (VTHC) (2006) *The view from the frontline: A report on the experience of OHS reps*, Melbourne, Victorian Trades Hall Council.

Vosko, L. (2007) 'Gendered labour market insecurities: Manifestations of precarious employment in different locations' in V. Shalla and W. Clement, (eds) *Work in Tumultuous Times: Critical Persepectives*, Montreal: McGill Queen's Press: 52–97.

Vogel, L. (1993) *Prevention at the workplace. An initial review of how the 1989 Community framework Directive is being implemented*, Brussels, TUTB.

Vogel, L. (2004) The community strategy at mid-term, *TUTB Newsletter*, 26: 17–30.

Vogel, L. (2008) 'Workers and safety representatives participation. The key to success in risk assessment', *Magazine of the European Agency for Health and Safety at Work*, 11: 6–8. Bilbao.

Walters, D. (1987) 'Health and safety and trade union workplace organisation: A case study in the printing industry,' *Industrial Relations Journal* (18) 1: 40–9.

Walters, D. (1996) 'Trade unions and the effectiveness of worker representation in health and safety in Britain,' *International Journal of Health Services*, 26: 625–41.

Walters, D. (2001) *Health and safety in small enterprises: European strategies for managing improvement*, Brussels, P.I.E. Peter Lang.

Walters, D. (2002) *Working safely in small enterprises in Europe: Towards a sustainable system for worker participation and representation*. Brussels, ETUC.

Walters, D. (2003) 'Workplace arrangements for OHS in the 21st century', National Research Centre for OHS regulation, Working paper 10, Canberra, The Australian National University.

Walters, D. (2006) 'One step forward, two steps back: Worker representation and health and safety in the United Kingdom', *International Journal of Health Services*, 36 (1): 87–111.

Walters, D. (ed.) (2002) *Regulating health and safety management in the European Union*, Brussels, Peter Lang.

Walters, D. and K. Frick (2000) 'Worker participation and the management of occupational health and safety: Reinforcing or conflicting strategies?', in K. Frick, P. L. Jensen, M. Quinlan and T. Wilthagen (eds) *Systematic Occupational Health and Safety Management – Perspectives on and International Development*, Oxford, Pergamon.

Walters, D. and Gourlay, S. (1990) *Statutory employee involvement in health and safety at the workplace: a report of the implementation and effectiveness of the safety representatives and safety committees regulations 1977*. London, Health and Safety Executive, Report 20/1990, London (accessed 1 May 2006), http://hesa.etui-rehs.org/uk/dossiers/dossier.asp?dos_pk=15.

Walters, D. and Kirby, P. (2003) *Training and action in health and safety*, London, TUC.

Walters, D., Kirby, P. and Daly, F. (2001) *The impact of trade union education and training in health and safety on the workplace activity of health and safety representatives*, London, Health and Safety Executive Research Report 321 (accessed 1 May 2006), http://www.hse.gov.uk/research/crr_pdf/2001/Crr01321.pdf.

Walters, D. and Nichols, T. (2007) *Worker representation and workplace health and safety*, London, Palgrave Macmillan.

Walters, D., Nichols, T., Connor, J., Tasiran Ali, C. and Cam, S. (2005) *The role and effectiveness of safety representatives in influencing workplace health and safety*, London, Health and Safety Executive, http://www.hse.gov.uk/research/rrpdf/rr363.pdf.

Walters, V. (1983), 'Occupational health and safety legislation in Ontario: An analysis of its origins and content' *Canadian Review of Sociology and Anthropology*, 20: 413–34.

Waxman, H. A. (2004) *A special interest case study: The chemical industry, the Bush Administration, and European efforts to regulate chemicals*, Washington, United States House of Representatives Committee on Government Reform – Minority Staff Special Investigations Division.

Wedderburn, K. W. (1980) 'Industrial relations and the courts', *Industrial Law Journal*, 9: 65–96.

Wikman, A. (forthcoming) 'Changes in power, influence and organisation' in S. Marklund, and A. Harenstam (eds), *The Dynamics of Organizations and Healthy Work*.

Wilthagen, T. (1994) 'Reflexive rationality in the regulation of occupational health and safety' in R. Rogowski and T. Wilthagen (eds), *Reflexive Labour Law*, Deventer, Kluwer.

Woolf, A. (1973) 'Robens report – The wrong approach?', *Industrial Law Journal*, 2 (1): 88–95.

Woolfson, C. (2004) *Regulation of the working environment in the new accession states of the enlarged European Union*, Brussels, TUTB.

Woolfson, C. (2005) 'Health and safety at work after enlargement: A new European strategy, or more of the same?' *Policy and Practice on Health and Safety*, 3 (2): 69–88.

Woolfson, C. (2006) 'Working environment and soft law in the post-communist new member states', *Journal of Common Market Studies*, 44 (1): 195–215.

Woolfson, C. and Beck, M. (2003) 'Occupational health and safety in transitional Lithuania', *Industrial Relations Journal*, 34 (3): 241–60.

Woolfson, C., Beck, M. and Šceponavicius A. (2003) 'Workplace health and safety in transitional Lithuania: A survey', *Policy and Practice in Health and Safety*, 1 (1): 59–83.

Woolfson, C. and Calite, D. (2008) 'Working Environment in the new EU Member State of Lithuania: Examining a "worst case" example', *Policy and Practice in Health and Safety*, 6 (1): 3–28.

Index